An Introduction to
Geotechnical Processes

An Introduction to Geotechnical Processes

JOHN WOODWARD

Spon Press
Taylor & Francis Group

LONDON AND NEW YORK

First published 2005
by Spon Press
2 Park Square, Milton Park, Abingdon, Oxon OX14 4RN

Simultaneously published in the USA and Canada
by Spon Press
270 Madison Ave, New York, NY 10016

Spon Press is an imprint of the Taylor & Francis Group

© 2005 John Woodward

Typeset in Helvetica by
Newgen Imaging Systems (P) Ltd, Chennai, India

Printed and bound in Great Britain by
Bell & Bain Ltd, Glasgow, Scotland

British Library Cataloguing in Publication Data
A catalogue record for this book is available from the British Library

Library of Congress Cataloging in Publication Data
Woodward, John, 1936–
    An introduction to geotechnical processes / John Woodward. – 1st ed.
      p. cm.
    Includes bibliographical references and index.
    1. Engineering geology. I. Title.
TA705. W66 2004
624.1'51–dc22   2004001287

ISBN 0–415–28645–X (hb: alk. paper)
ISBN 0–415–28646–8 (pb: alk. paper)

For Emily and George

PREFACE

Nowadays geotechnical processes are used in a range of civil engineering projects to assist in construction by stabilising ground or to improve the load-carrying capacity of foundations. Although numerous specialist publications and state-of-the-art dissertations exist on the wide variety of topics involved, it is difficult to find one text dedicated solely to bringing together for the non-specialist the most commonly used ground treatment techniques. I have therefore endeavoured to address this gap and provide in this book a concise and basic compendium of the essential practical information for as wide a readership as possible at minimal cost. It is not a comprehensive design manual, but rather a review of practical answers to questions raised by the non-specialist when faced with making a decision on ground treatment – 'What bearing capacity can I expect following treatment?', 'Can this soil be frozen and what about heave?', 'Will dynamic compaction cause problems in the adjacent buildings?', 'Must I spend all this money on ground investigation?' and so on.

In offering guidance, as a practitioner rather than a theorist, on such diverse subjects, I have had in mind the general civil and structural engineer who would like to be better informed when turning to geotechnical specialists for assistance in dealing with difficult ground conditions; also the developers, estimators and procurement personnel who need to be familiar with the concepts when dealing with proposals from specialists and pricing such works. In addition I hope this book will provide those civil engineering students and engineering geologists who may be considering specialising in this interesting area of construction with an introduction to practical foundation engineering.

'Geotechnical processes', for the purpose of this book, cover ground treatments and improvements, generally undertaken by specialists, which use:

- extraction and exclusion of groundwater – typically by wells and cut-offs
- compaction techniques – by vibration and dynamic pounding and
- drilling and grouting techniques – by injecting self-setting fluids into ground

to stabilise ground, improve the support of structures and control settlement.

The complex subject of piling has not been considered as a geotechnical process for this *Introduction*, except briefly where piles are used to form cut-off barriers. Reinforcing of soils using geofabrics has not been included as this is usually undertaken directly by the civil engineering contractor.

The text is based on the notes provided for a regular series of lectures given by the author to post-graduate engineering geologists at Imperial College, London, and to practising civil engineers over the past 10 years. These lectures were in turn based on data collected, derived and researched from many sources during my career as a geotechnical contractor and consultant, and these sources are acknowledged in the text and fully referenced.

The format is a simple double-page layout (reduced or expanded to suit the scope being considered), so that each of the processes considered will stand alone for reference. This has inevitably led to retaining the lecture note style, rather than using a narrative form, in order to give a broad and accessible view of what is currently available. Each process is described and illustrated with cases and newly drawn diagrams, many based on the author's experience, under common headings ranging from the principle of the process and its applications to construction methods, monitoring and plant capabilities. There is some deliberate repetition of ground investigation requirements to remind readers of the importance of accurate and appropriate data. Health and safety precautions are also highlighted. Cross-references between sections are given in the form (4.1).

Although every effort has been made to use reliable sources and to verify the information given, the ranges and limitations quoted for the various parameters, conditions and improvements possible using the techniques discussed are unavoidably fairly broad. Clearly no guarantees can be given here as so much depends on the specific geology, context and application, and particularly on practical experience. Expert advice should always be sought. The author is indebted to Dr Bryan Skipp for advice and comments on the text, Dr Stephen Thomas of OGI (Oxford Geotechnica International) for providing finite element analysis for solving fluid flow in soils, Mr Ian Higginbottom for reviewing the text and diagrams and providing geological notes, Mr Peter Constantine for reviewing the soil mechanics and Mr Richard Miles of SeaMark Systems for an update on offshore grouting.

Thanks are due to the following companies and individuals for granting permission to use photographs and copyright material:

Bachy-Soletanche Ltd – photographs of reinforced earth in 9.4 and rotating cutters in 11.1.

Cementation Foundations Skanska Ltd – photographs of piled walls in 5.6 and Vibwall in 5.9.

Ischebeck-Titan – photographs of anchored sheet pile wall in 5.5 and abseiling in 9.3.

Dr D P McNicholl – diagram of slope drainage in 4.1.

© Pearson Education Limited 1983, 2000, reprinted by permission of Pearson Education Limited. R Whitlow, *Basic Soil Mechanics* (1983, 2000) – diagrams of shrinking/swelling of clays in 1.6 and groundwater conditions in 3.1.

Dipl.-Ing. H J Priebe – diagram of stone columns in 6.4.

SeaMark Systems Limited – photograph of the large jet mixer in 10.1 and diagram of mixer in 11.2.

Sireg SpA – photograph of spiling in 9.5.

© Spon Press – stereogram in 1.4; depth of treatment by dynamic compaction in 6.5; grout pressure diagram in 7.3; jet grout column diameter in 7.6; diagram of old mine workings in 8.1; karst features in 8.3.

© Swets and Zeitlinger Publishers – groutability index in 7.1 and arrangement of grout circuits in 10.2.

Thomas Telford Limited – diagrams of Hong Kong diaphragm wall and cut-off at Sizewell B in 5.4 and compensation grouting in 7.4.

© John Wiley & Sons, Inc., 1974. R B Peck, W E Hanson and T H Thornburn, *Foundation Engineering* (1974) – extract from graph of SPT value versus relative density in 1.2.

© John Wiley & Sons, Inc., 1992. J P Powers, *Construction Dewatering – New Methods and Applications* (1992) – extract from eductor diagram in 4.5.

George Wimpey UK Limited – photographs in 1.7, 3.5, 4.1, 4.3, 4.5, 5.3, 6.3, 7.1, 8.3, 9.3 and 11.1.

Zschokle Services AG, Dietlikon – photographs of spraying shotcrete in 3.6, the shotcrete wall in 5.6, the jet grouted support in 7.6, a jet grout monitor and grab in 11.1.

Extracts from BS 8081: 1989 are reproduced with the permission of BSI under license number 2003DH0301. British Standards can be obtained from BSI Customer Services, 389 Chiswick High Road, London, W4 4AL. Tel +44 (0)20 8996 9001. email: cservices@bsi-global.com.

J. C. W.
Princes Risborough 2004

SYMBOLS AND ABBREVIATIONS USED

The unit for stress and pressure used in this book is kN/m^2 (kilonewtons per square metre) or where appropriate (e.g. for unconfined compressive strength) N/mm^2 (newtons per square millimetre) and MN/m^2 (meganewtons per square metre). Also used is the bar unit of pressure when referring to injections and pump capabilities.

$1\ kN/m^2 = 1\ kPa$ (kilopascal in SI units)
$1\ N/mm^2 = 1\ MN/m^2 = 1\ MPa$ (megapascal)
$1\ bar = 100\ kN/m^2$

a	width of a flow net square
A	area
A_c	area of stone column
A–D	weathering grades for chalk
amp	ampere
ASTM	American Society for Testing and Materials
A_w	drain cross-section area
b	breadth of sliding black
B	base area of temper
BH	borehole
c, c_u, c'	cohesion (shear strength), undrained shear strength, drained shear strength
CCTV	closed-circuit television
CDM	Construction, Design and Management Regulations 1994
CE	Civil Engineering grade of bentonite
CFA	continuous flight auger
CFRP	carbon fibre reinforced polymer
c_h	coefficient of consolidation (horizontal)
COSSH	Control of Substances Hazardous to Health (Regulations)
cp	centipoise
CPT	cone penetration test
c_v	coefficient of consolidation
CWS	continuous water-stop
c/c	centre to centre
D	diameter, or depth
dB A	decibel
DC	direct current
d_e	equivalent diameter
dia.	diameter
DIN	Deutsches Institut für Normung
DTH	down-the-hole (hammer drill)
d_w	diameter of vertical drain
D_x	diameter than which x percent of soil particles are finer
E, E_m	elastic modulus, deformation modulus of rock mass
EA	Environment Agency (UK)
EFT	a type of 'Vibwall'
F	factor of safety
Fe	iron
f_{ult}	ultimate rock/grout bond stress
g	acceleration produced by gravity
GGBS	ground granular blast-furnace slag
GL	ground lovol
GWL	groundwater level
h, H	hydraulic head, or height
h	hour
h_a	groundwater head at point a
HDPE	high-density polyethylene
HP	high-pressure
HWST	high-water spring tide
i	hydraulic gradient
I–V	weathering grades for rock
j	rock mass factor
k	coefficient of permeability (hydraulic conductivity)
K	factor of soil resistance

l	litre
L	length of fixed ground anchor
LN	liquid nitrogen
l_w	wetted length of well screen
m	metre
MDPE	medium-density polyethylene
mg	milligram
min	minute
mm	millimetre
m_v	coefficient of volume compressibility
n	ground anchor installation factor, or number of wells, or ground improvement factor
N	standard penetration test, blows per 300 mm
N, N_R	groutability ratios (soil and rock)
NATM	New Austrian Tunnelling Method
N_f, N_p	number of flow lines and pressure drops, respectively
Nm	torque, newton metre (or kNm, kilonewton metre)
No	number of
NVQ	National Vocational Qualification
OD	outside diameter or Ordance Datum
OGL	original ground level
OP(C)	ordinary Portland (cement)
PFA	pulverised fuel ash
PVC	polyvinyl chloride
PVD	prefabricated vertical drain
Q	rock quality rating (Norwegian system)
Q, q	quantity (yield) of water
q_w	quantity of water (well yield)
r	well radius
R	radius of cone of depression, or resistance to sliding
r_a	radius of cone depression at point a
RAMP	Risk Analysis and Management for Projects
RC	reinforced concrete
RH	rockhead
RHOPC	rapid-hardening ordinary Portland cement
RMR	rock mass rating
ROV	remotely operated vehicle
r.p.m.	revolutions per minute, rotation speed
RQD	rock quality designation
s	second
s	spacing of vertical drains
s_c	settlement with stone columns
S	specific yield
SPT	standard penetration test
t, tonne	metric tonne
T	transmissivity, or tension in ground anchor
TAM	tube à manchette
TBM	tunnel boring machine
TCR	total core recovery
T_f	pull-out resistance of ground anchor
u, u_e	porewater pressure, excess porewater pressure
U	uniformity coefficient
UCS	unconfined compressive strength
UKAS	UK Accreditation Service (for laboratories)

uPVC	unplasticised polyvinyl chloride	α	angle of slope
V	volt	β	angle of rock anchor
W	watt	γ	effective unit weight
wt	weight	γ_w	density of water
WT	water table	ϕ, ϕ_u, ϕ'	angle of internal friction, undrained angle of internal friction, drained angle of internal friction
W	weight		
z	depth	μm	micron
3D	three-dimensional	σ	normal stress
3ph	three phase (electrical supply)	σ'	effective normal stress
°C	degree Celsius	τ	shear strength

Abbreviations used for literature citations are defined in the References.

CONTENTS

INTRODUCTION

The newcomer to geotechnical works will soon be aware that the science and technologies involved are full of uncertainties which may not be readily quantified. This can make it difficult to explain to clients and non-specialists precisely why ground treatment is being proposed, what the objectives are and how they are to be achieved and verified. The problem stems from the variable nature of the soil and rock on which all structures sit. The structural engineer can define dimensions, strength of construction materials and the loadings in a building to a fair degree of accuracy, but the geotechnical engineer is faced with having to unravel and evaluate the complex 'soil–structure interaction'. The uncertainties range from knowing whether the scientific assumptions and models which can be reasonably applied to the combined structure–foundation–subsoil are right, to the largely random nature of geological features and the dispersion of soil properties by chance.

The geotechnical engineer's judgements will therefore be at the heart of the interpretation of ground conditions and decisions on methods of foundation treatment – and usually all there is to work with are the findings from a few boreholes. However, in order to provide the necessary degree of confidence in design recommendations for ground treatment, the geotechnical engineer will not base judgements solely on such limited information. Uncertainties in the findings can be tested against experience of the geology generally and expectations for material properties using the extensive database and robust theoretical models now available to extend knowledge of the site.

Despite such evaluations and investigations to look for potential flaws, adverse ground conditions which cannot be fully resolved by current foundation engineering theory and practice remain a possibility. In these situations the application of systematised aids to decision-making, such as the interactive 'observational method', to direct the judicial use of geotechnical processes can be crucial for project realisation. Here the geotechnical engineer will consider the risks and probabilities of different construction scenarios and variations in material properties arising, then set suitable but not overly conservative design boundaries within which the geotechnical works can be assessed, modified and verified on a regular basis during construction. Such ongoing designed modifications can cause budgeting complexities for the client, but properly executed the method gives added security for the works and is cost-effective.

An alternative means of reducing uncertainties prior to embarking on construction is the 'probabilistic method' (Whitman, 2000), which attempts to deal with the consequences of spatial variability of parameters, the adequacy of testing, variations in strata and levels, and many other potential errors and scatter in the data. Use is made of 'search' theory, reliability theory and statistical and probability analyses to express the risks numerically to assist in early decision-making for geotechnical solutions.

In contracting terms, the problems of 'unforeseen' adverse ground conditions arising from uncertainties due to limited or highly variable data have to be accounted for: should the risk and cost of dealing with them fall to the engineer, the contractor, the client, the insurers or be shared on some basis? Designing an investigation specifically to produce parameters for a particular process can be done only after the general conditions (the 'ground model') have been ascertained. If this iterative sequence is not followed then poor or inadequate initial data can waste time in down-the-line interpretation and result in contractual claims for unforeseen conditions.

But it has to be recognised that in any particular case the engineering judgement may be that further complex investigations would be costly and possibly unrewarding – leaving open the question of what may be deemed unforeseeable.

It is necessary when considering reasonable foreseeability to examine how random the conditions are and the chances of variation in properties. For example, the chances of finding flints in the Upper Chalk are reasonably certain, but no flints were recorded in a particular ground investigation for a tunnel in the Chalk. The tunnelling machine ran into a bed of flints which damaged the tunnelling machine, and the contractor claimed costs for overcoming unforeseen ground conditions. Is this reasonable? How many more boreholes should have been drilled to recover pieces of flint? What should the geotechnical engineer do to avoid the need for such claims to be made? The presence of fissures, faults, ancient landslides, buried erosion channels, etc. may be so random and unusual that conditions result which even the experienced geotechnical engineer cannot reasonably foresee. Essentially, the expert specialist should be able to assess and grade the quality of data gathered for foundation design, advise the interested parties on how best to apply the data and warn of the uncertainties which remain. The non-specialist client and engineer should ensure that the geotechnical specialist is involved from the start, be aware of the consequences of having to infer design parameters from limited data and use risk assessment and contractual terms to mitigate the effects of the uncertainties.

British Standards and Codes of Practice are referenced throughout this book, as geotechnical engineers will be faced with compliance in the commercial world. These documents have been criticised for not necessarily being 'best practice', for being out of date within a few years and for not dealing sufficiently with the basic theoretical principles. They are not intended to be a substitute for a sound background in the theories of soil and rock mechanics which must be the foundation of geotechnical expertise. But the methods and procedures recommended are based on practical experience and sound empirical data to produce designs and constructions in which the various parties to a contract can have confidence. The Standards should therefore be followed unless there is good quality theoretical and practical evidence to show that an alternative procedure or method will be more beneficial and progress the technology.

Eurocodes have not been specifically referenced at this time in view of the ongoing development of drafts. But as final publication is at hand, it is recommended that all new geotechnical designs are considered against the requirements in Eurocode 7 parts 1–3, out for comment (Simpson and Driscoll, 1998; Hope, 2003).

There is a wide range of regulations and approved codes of practice with legal standing under the Health and Safety at Work Act 1974 which can impact geotechnical works – the Control of Substances Hazardous to Health (COSSH) Regulations, the Ionising Substances Regulations, the Construction (Lifting Operations) Regulations and more. These documents can be complex; they should be studied with care and professional advice should be sought.

Most of the works described in this book will be carried out by specialist subcontractors under contracts awarded and administered by civil engineering and building contractors. Occasionally the client may 'nominate' a subcontractor to carry out the work because of the firm's expertise in a process or familiarity with the site conditions. The ground investigation specialist will most ofton bo appointed by the client, with the scope of work designed and supervised by the Engineer (the client's agent), although the specialist will design any temporary works necessary. The design, execution and performance of ground treatment work is frequently the sole responsibility of the specialist, with the requirement to satisfy the Engineer that all treated ground has attained the required degree of improvement. Whatever the form of contract, the obligations and liabilities for carrying out grotechnical works should be clearly set out by the parties in accordance with appropriate professional advice.

The decision to apply geotechnical processes to improve soil and rock conditions is therefore based on co-operation not only between the structural and geo-technical engineers, but also between the geotechnical engineers and the client, contractor and the other technical and commercial advisers. Communications between the parties are vital to the understanding of the risks involved and to ensuring that these can be controlled by appropriate design, construction and monitoring.

This nutshell guide outlining some basic practical matters involved in ground treatment is aimed at fostering good communications.

1 GROUND INVESTIGATION

1.1 General Requirements

The major technical and financial risks for many civil engineering works lie in the ground. The 'Site Investigation in Construction' series produced by the Institution of Civil Engineers (ICE, 1993a) provides guidance on planning, procurement and specifications for ground investigations which will help to minimise risk. It gives the warning that 'you pay for site investigation whether you have one or not' – meaning apparent savings on investigation at the start of a project may well lead to considerable additional later costs in coping with 'unforeseen' ground conditions.

It is usually necessary to investigate more than the geological and geotechnical conditions on any site in order to produce a meaningful risk assessment and report on the proposals for civil engineering and building foundations.

This general 'site investigation' should determine the physical characteristics of the site as they affect design and construction of the works and the stability of neighbouring structures. Specialist input may be needed from appropriate sources in order to cover the following aspects:

- land-ownership and access
- previous uses of the site and environmental issues
- the condition of existing adjacent buildings, basements, underground services and overhead cables
- demolition/disposal requirements.

The specialist 'ground investigation' to explore and record the locations and characteristics of soils, rocks and ground conditions will be the main feature of the report, providing the geotechnical parameters needed for the design and construction. Objectives will be to:

- obtain appropriate data on the site and ground conditions to ensure the site is suitable for the proposed structures and
- determine whether any special measures are necessary to support the structures to allow construction to proceed safely and/or to improve the ground.

These objectives raise two dilemmas for the client and engineers:

- When during the ground investigation can a decision be taken to assess the need for a particular ground treatment or improvement which will allow foundations to be built economically? The answer lies with the 'phased' investigation approach.
- Whatever the extent of ground investigation it cannot compare with the information which becomes available during construction. The answer lies with the 'observational method'.

The cost of ground investigation depends on the size of the project, and no general rules can be offered. Typically expenditure on investigations for buildings is low, say, 0.2 per cent of the construction cost, and may be worth very little in terms of value for money and risk reduction if it has not been properly planned and executed. At the other end of the scale, a large dam will require 3–5 per cent of the total construction cost to be spent on investigation. In such cases the cost of onsite trials for, say, dewatering or a grout curtain may be over 50 per cent of the total investigation cost in order to provide useful data to reduce sensitive construction times.

But note that 'forensic' ground investigation into causes of foundation failure will usually be much more detailed and costly than investigation for new work.

Phased ground investigations are particularly important when assessing a site which may contain difficult ground conditions requiring some form of treatment or improvement.

Experienced geotechnical engineers or engineering geologists should develop the investigation phases appropriate to the complexity and size of the project in conjunction with the designers. As the information from each phase is obtained it should be used on an iterative basis to check and modify the proposed design and construction methods and instigate additional fieldwork and reports.

A comprehensive investigation will involve:

- desk study
- walk-over survey
- preparation of a conceptual model of the ground
- design of the preliminary ground investigation based on the model
- preliminary investigation and laboratory testing
- assessment of results and preliminary report as needed
- design and implementation of further specific investigations and onsite trials
- final report and recommendations
- monitoring and observations during construction
- amendment of requirements and procedures as needed.

Desk study to review existing records in conjunction with relevant specialists:

- topographical maps
- geological maps and regional literature
- mining records and plans
- previous site investigation and hydrological reports
- past experience of ground treatments successfully applied in the area
- aerial photographs, infra-red photographs
- geographic information systems
- environmental issues, landfill/contaminated land
- noise and vibration studies.

Walk-over survey and reconnaissance (BRE Digest 348, 1989) to inspect and assess:

- geomorphology – the general shape of the ground within and around the proposed site correlated with the geological map, the aerial survey and local strata exposures (Fookes, 1997; Griffiths, 2001)
- groundwater springs, existing adjacent water courses, tidal water and standing water
- patterns of drainage, potential recharge and source
- condition of existing adjacent structures
- presence/absence of clay/shrinkable clay
- likely presence of alluvium or peat
- influence of trees
- made ground and fill, potential contaminated soil
- undermining – sinkholes/subsidence, spoil tips
- landslides, slope instability, depressions, erosion features, etc.

A 'conceptual ground model' is developed from the desk study and reconnaissance and will identify initially the general nature of the ground beneath the site – the likely strata

The intention should be to produce three-dimensional geological, hydrogeological and geotechnical data on the soils and rock beneath (and possibly adjacent to) the site, including manmade layers and features, in an accessible form which can be used by all members of the project design team.

and depths and groundwater conditions which need to be investigated. The model will be updated as the results of the investigation become available and are interpreted in relation to the structures and excavations being considered – particularly with reference to risk assessment (1.5).

The preliminary ground investigation will be based on the desk study and surveys. It should be designed and costed in conjunction with the client and design engineers using the recommendations, procedures and techniques given in BS 5930: 1999 'Code of Practice for Site Investigations', Clayton *et al.* (1995) and Joyce (1982), and include allowances for the following:

Fieldwork in the appropriate number of boreholes to the recommended depths using overburden and rotary drilling methods, trial pits and trenches to provide:

- geological and hydrogeological data on the depth of ground which is influenced by structural loading and may need improvement by geotechnical processes
- information on depth to rock, condition and shape of rockhead, 'rippability', fissuring, voids and solution features
- adequate samples of soil and rock for logging strata and laboratory testing
- *in situ* testing data – penetration tests, vane tests, pressuremeter tests, etc.
- data on groundwater levels, perched water, artesian and spring sources, variations in water levels due to seasonal conditions or tidal influence using piezometers and standpipes
- *in situ* permeability data and pumping tests
- geophysical survey data, correlated with boreholes
- an updated conceptual ground model.

Typical depths of boreholes should include all strata likely to be affected by the structural load and construction operations, e.g.:

- shallow foundations – at least 1.5 times the width of the base below the anticipated formation
- piles/pile groups – at least below the expected depth of stressed soil
- retaining walls – between 3 and 4 times the retained height below the founding level
- groundwater control – 1.5–2 times the depth of excavation, depth to safe aquiclude, depth of significant rock fissures
- ground anchors – at least 1.5 times the depth of the anchoring block of strata.

If rockhead is located within these depths, prove rock for 3 m – more where glacial till exists over rock and where solution features are possible.
The depths of holes for tunnelling and for construction over old mine workings must relate to the depth of stressed ground.

Number and location of boreholes depend on the likely strata and the type of structure, typically:

- 10–30 m centres for buildings
- 300 m centres for roads and tunnels; closer to prove features (buried channels, fissures and anomalies) and for each structure
- 100–500 m for large hydraulic structures e.g. channels; closer for dams and weirs.

Holes should be sampled at regular depths in soil and cored in rock.
The investigation should be flexible to allow for additional work as needed.

Laboratory testing of soil and rock should be as recommended in BS 5930: 1999 and in the Association of Geotechnical and Geoenvironmental Specialists' Guide (AGS, 1998), depending on the structures proposed. Testing should be carried out in accordance with BS 1377: 1990 'Methods of Test for Soils for Civil Engineering Purposes' and as expanded by Head (1986). The test results should be used to update the conceptual ground model. Note that variability of the soil and rock will produce a scatter of test results – if statistical analysis is to be made of large amounts of data, care must be taken not to ignore any results which are outside the distribution curve but may be crucial to stability (Trenter, 2003). Also:

- laminated soil may require special procedures – careful examination is advised
- samples should be sufficiently large and representative
- descriptions of soil must conform to the standard engineering classification
- testing laboratories should be accredited by the UK Accreditation Service (UKAS) or similar.

Supervisors and personnel must have appropriate experience and specialised knowledge of the work being undertaken, demonstrated by recognised current training and accreditation certificates:

- Supervising engineers should be trained as advised in the ICE 'Site Investigation in Construction' series (1993a), document 2 'Planning, Procurement and Quality Management'.
- Drillers should be accredited by National Vocational Qualifications (NVQs) as assessed by the British Drilling Association (BDA).

Health and Safety requirements on all sites are covered in a wide variety of statutory and advisory documents, and must be implemented and supervised by experienced personnel. Publications from the HSE, BDA (2002), AGS (2002), ICE (1993b) and by M R Harris *et al.* (1995) provide comprehensive relevant references covering e.g. laboratories, contaminated land, old mine workings, work over water, etc.
Plant and equipment are summarised in 1.7 below.

Specific geotechnical data requirements will be developed from the preliminary model and report, and additional investigations will target areas where anomalies and difficult ground conditions have to be resolved. Some relevant stability parameters are given below (1.2) and in the following chapters for each geotechnical process considered.

The ground investigation report should establish clearly the soil profile and rock stratigraphy (Powell, 1998), the groundwater conditions for the conceptual ground model and reliable strength of materials data presented to the design engineer. Engineers other than those producing the factual data for the ground model may be called on to provide expert interpretations of the ground conditions and safe construction and operation of the project.
The report should follow the client's brief strictly, and should be unambiguous and accurate. Recommendations must logically follow the factual findings. BS 5930: 1999 provides guidance on preparing the final report, including the presentation and interpretation of data. The AGS document 'Electronic Transfer of Geotechnical Data from Geotechnical Investigations' (2000) provides a format and conventions for standard presentation and transmission of factual information in the report. Where a 'comprehensive' site investigation report involving input from other specialists is not provided to the client, the ground investigation report should include the relevant information on site usage, services, environment, etc.

1.2 Stability Parameters

Geotechnical design for foundations, piling, ground treatment, groundwater control, slope stability, tunnelling, etc. requires the assessment of fundamental stability parameters both during the phased ground investigation and for quality control during construction.

In order to produce appropriate parameters, it is useful for the geotechnical engineer, in conjunction with the structural designer, to rank and cost a selection of foundation methods suitable for the conditions shown in the conceptual ground model before finalising any supplementary investigation.

The cost of such additional field and laboratory testing and site trials will usually pay dividends by improving judgement and decision-making.

> **Note**: even the 'best' investigation will leave parts of the site unexplored.

The geotechnical engineer must be familiar with design and construction methods in order to liaise effectively with the design engineer and enable his or her knowledge of soil mechanics and geology to be applied with proper judgement to a specific problem. In particular, the geotechnical engineer must be aware of the limitations of geotechnical processes for improving ground conditions and advise the design engineer accordingly.

Project data which the geotechnical engineer will need in addition to the 'interpretative' ground investigation report and the conceptual ground model to examine ground treatment methods and determine the relevant foundation stability parameters include:

- structures proposed, with estimated applied loads
- foundation excavation areas and depths
- likely method and rate of foundation excavation
- proximity of adjacent structures
- limitations on settlement and heave
- limitations on noise and vibrations
- preliminary construction programme
- availability of excavation plant.

The quality of geotechnical data is important if the geotechnical engineer is to make use of advances in soil mechanics theory and apply them to achieve cost-effective foundations. For example:

- penetration tests must not be performed in boreholes where 'blowing' of the material is occurring
- time must be allowed to record groundwater levels in boreholes (including seasonal variations) and for laboratory tests for shear strength
- truly 'undisturbed' samples do not exist – but with care this need not unduly affect the test result
- good correlation should be ensured between laboratory tests results, *in situ* tests and field observations
- if appropriate, more refined equipment can be used (thin wall and piston samplers).

SHEAR STRENGTH

The **shear stress** in a soil resulting from the overburden and applied load is resisted by contact between the solid particles. The **shear strength** of soil (τ) is the limiting value of shear stress which a soil can support before it yields. This is expressed as a function of the two shear strength parameters, which depend on the soil particle sizes, c and ϕ (i.e. cohesion and angle of internal friction, respectively) and normal stress (σ):

$$\tau = c + \sigma \tan \phi.$$

The shear strength of coarse-grained soil depends on the degree of interlocking between the particles and is therefore a function of the relative density (e.g. *N-value*) of the soil. The shear strength of cohesive soils ('fine soils – clays and silts which can sustain suction when saturated and unconfined' as BS 5930: 1999) depends on the chemical bonding of the mineral platelets and porewater pressure.

Triaxial load cells (BS 1377: 1990) are the most widely used laboratory shear strength test for fine soils, allowing control of drainage conditions and porewater measurements. The test specimen is obtained from carefully extruded 'undisturbed' tube samples.

Cohesive soil will have a 'peak' strength, which will decline with time to 'residual' strength as the soil fabric is re-aligned under applied pressure. Both cohesion and friction angle are usually reduced in this residual state.

The test procedure will be varied depending on the degree of saturation at the stage of construction/loading being considered:

- **unconsolidated, undrained**, with no drainage allowed – which gives an initial *in situ* condition
- **consolidated, undrained**, with drainage allowed until consolidation is complete, then continued undrained with porewater pressure measurements
- **drained**, with drainage allowed throughout the consolidation and shearing stages.

The undrained strength can be given in total stress terms (c_u and ϕ_u), and the drained strength in 'effective strength' terms (c' and ϕ') (3.1).

The tests and interpretation of results are complex and should be undertaken by experienced engineers and technicians.

Field vane tests are used for *in situ* undrained shear strength measurement in low-strength saturated clays.

The shear strength of rock is defined, as above, with two parameters, cohesion and angle of internal friction. Because of the much higher strength of rock, the triaxial test is not used except for weak rock, e.g. mudrocks and chalk.

Unconfined compressive strength (UCS) is the most common laboratory test on a prepared cube or cylinder of intact rock and is used generally to define rock strength. Shear strength may be estimated from the UCS as UCS/6 in strong intact rock.

Laboratory shear box tests on small specimens are used to determine shear resistance of discontinuities – which is more critical than intact shear strength.

In situ shear tests require a soil or rock sample to be prepared in a pit or tunnel with a means of applying and measuring vertical and horizontal load and movement.

ELASTIC/SHEAR MODULUS

The relationship between stress and strain under particular porewater pressures and loading will indicate whether the soil or rock will be 'brittle' (a clear breaking or shearing point), 'elastic' (shape recoverable on unloading) or 'plastic' (shape not recoverable on unloading). The tangent or secant slope of the stress/strain curve is designated the elastic modulus (E) and can be determined in the laboratory using the triaxial apparatus under various drainage conditions or UCS test for rock.

This modulus is important in determining settlement due to loading of the ground.

Plate loading tests to produce the stress/strain curve in trial pits or large-diameter boreholes avoid difficulties with sample disturbance.

Pressuremeters (1.5) are used extensively for measuring the shear modulus and stiffness of most soils and weak rocks *in situ*.

CONSOLIDATION

The estimation of consolidation and settlement of a soil under load requires determination of several soil mechanics parameters and is also a vital part of groundwater control and ground improvement.

Laboratory oedometer tests (one-dimensional loading) (BS 1377: 1990) are used to determine the coefficient of volume compressibility (m_v) of clays and silts, from which the coefficient of consolidation (c_v) is obtained. These two coefficients enable the magnitude and rate of consolidation under the full structural applied load to be estimated. However, as the standard oedometer sample is small (75 mm diameter, 20 mm thickness) these calculations may not be representative of the *in situ* conditions. Examination of clay samples (for fine sand partings or thin lenses) is necessary as the soil fabric will influence both the permeability and consolidation of clay soils. For example, anisotropy in the specimen will produce different vertical and horizontal consolidation coefficients (by a factor of 2–6 for clays) – which is significant when considering ways of accelerating the settlement of clay soils by using vertical drains.

The use of larger specimens for testing in a Rowe cell, to give more realistic values, can improve the consolidation estimates.

Atterberg limit tests for soil consistency may be used to provide an empirical relationship for the 'compression index' (Bowles, 1988).

Piezocone tests, which induce high excess porewater pressure during driving of the cone into fine soils, can provide an estimate of the *in situ* c_v by observing the rate of dissipation during breaks in driving.

Field loading tests to observe settlement of soil under trial embankments to simulate the full-scale structure are also used to give consolidation and settlement data.

SETTLEMENT

The prevention of settlement, particularly differential settlement, is the reason for most ground treatment. This complex process is caused by:

- *compaction* when applied load, removal of groundwater or vibration reduces the volume occupied by the soil particles – coarse sands and gravels are most susceptible
- *consolidation* when porewater is squeezed out under applied load – fine soils and clays are most susceptible
- *erosion* when fine material is removed from the soil fabric, leaving voids which collapse under load or self-weight over time.

Because of the complexity of the interaction between soil stratification (thickness, compressibility, permeability, etc.), anisotropy, porewater pressure and the applied load, absolute values of settlement are difficult to predict. To obtain an estimate, each layer has to be considered separately at various stages of construction (and service life of the structure) when assessing likely settlement (Tomlinson, 2001).

Application of load to, or removal of water from, coarse saturated soils can cause immediate settlement, whereas settlement of finer silts and clays will be longer term. Shrinkage of 'expansive' clays as they dry out is a major cause of settlement (1.6).

If the settlement is uniform, the effects on the structure being supported will be minimal (but note that services entering the building may be ruptured).

Differential settlement will be the structural designer's main cause for concern. Empirical limits for settlement have been set depending on the type of structure and impact on appearance, serviceability and stability (Tomlinson, 2001).

The designer will first consider structural means of avoiding differential settlement – rafts, piles, etc. – but pretreatment of ground may be necessary – e.g. dynamic compaction or drainage for an economic foundation.

Where groundwater lowering is considered, 'drawdown' (3.2) under adjacent structures must not be allowed to cause settlement as a result of either changed effective stress and piezometric level or erosion.

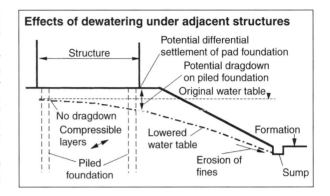

RELATIVE DENSITY/COMPACTION

Relative density is a measure of the density of granular material compacted mechanically in the field in relation to laboratory compaction or an indication of the state of compaction of *in situ* soil – indirectly indicating angle of friction, bearing capacity, potential for further compaction, permeation by grout, etc.

Standard penetration tests (SPTs) are the main method of assessment in granular soil, performed by driving a cone or 'split spoon' sampler (BS 1377: 1990) into a borehole to provide the *N-value*, which characterises the relative density for evaluating geotechnical processes.

Corrections are used to give *N-values* in relation to effective overburden pressure. Corrections to the field results are essential if non-standard equipment is used for the test in order to obtain the standard *N-values* (Clayton, 1993).

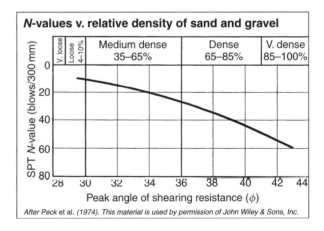

Electrical cone penetrometers are driven into a borehole, and the end (cone) resistance and sleeve resistance are measured to produce immediate readouts for sands and clays (Lunne, 1997).

Pressuremeters (1.5) are also used widely for tests in specific soils and fills.

Field compaction is measured by various sand or water 'replacement' methods (BS 5930: 1999).

1.3 Groundwater Investigations

At the start of major investigations it is unlikely that a specific geotechnical process will be contemplated; this can only be considered on the basis of the initial conditions found and the structures required.

Permeability and ground water levels, however, are two essential parameters for most conceptual models, not only when groundwater control has to be considered.

PERMEABILITY

Soils are permeable to a widely varying degree depending on the soil fabric, but for practical purposes a soil is considered impermeable when transmission of groundwater is so slow as to be negligible. Differences in vertical and horizontal permeability (anisotropy) will affect drainage and 'groutability'.

Intact rock core permeability can be several orders of magnitude lower than the overall *in situ* permeability, which is dependent on:

- the stratification – with the permeability along the bedding usually several times greater than that across the bedding (typically 5:1 and sometimes much greater)
- the degree of fissuring (width, continuity, infilling of discontinuities)
- solution features e.g. cavities.

Laboratory tests on rock cores taken from vertical boreholes will usually pertain to vertical permeability, whereas the *in situ* test of the rock stratum from which the core was taken will pertain to mass horizontal permeability.

Limestones and other soluble strata (e.g. gypsum) will exhibit solution and bedding features producing very variable permeabilities. Yields when dewatering, and grout-takes when forming cut-offs, will vary accordingly, from very high to negligible. The permeability of intact sandstones is more predictable and ranges from high (e.g. Sherwood Sandstone) to medium.

needed when comparing the results from the range of typical tests summarised below. The formulae used to calculate permeability should always be reported.

Selection of the method of testing must be relevant to the groundwater control or ground improvement systems being considered for construction; guidance and procedures are given in BS 5930: 1999, but many authors have produced references to refine testing methods (Preene *et al.*, 2000).

Field permeability tests should be carried out after sufficient data have been collected on the stratification and hydrogeology of the site. The records of tests should be presented in the final ground investigation report, but in any event must be retained for the duration of the project in order to assist in solving problems which may arise during and after construction.

Pumping tests using wells are the most reliable methods for determining mass permeability for groundwater control, particularly for large projects and complex geology – but they are the most expensive. The tests are carried out in either partially or fully penetrating wells (3.2). Pumping tests must:

- model the proposed method of groundwater control
- be monitored with adequate observation points to record the lowering of the water table – the 'drawdown' – and the recovery.

In addition to measuring rate of extraction and drawdown, it is useful for future construction to record:

- effects on adjacent structures and water courses
- recharge from adjacent water sources (check the use of tracer dyes with the Environment Agency)
- chemical analysis of groundwater (to assess future discharge disposal route).

Borehole permeability tests, which can target individual strata, are preferable for estimating *in situ* permeability when assessing grouting applications to exclude water and for ground improvement. Permeability estimates from borehole tests can be compromised by:

- not cleaning the borehole pocket before making the test
- clogging of the pocket, particularly in 'falling head' and 'constant head' tests
- loosening of the soil in 'rising head' tests or by drilling technique
- testing a small volume of soil in the borehole pocket
- thin layers of clay and silt in the formation
- anisotropy.

Use the maximum estimate of permeability obtained; the most reliable results are for coarse soils.

Tests in standpipes and porous pot piezometers can give good *in situ* permeability estimates in coarse soils, and hydraulic piezometers are satisfactory in fine soils; pneumatic or electrical piezometers are not suitable.

Borehole packer tests at constant head (a single packer as drilling progresses – 'descending' test – or double packer from the bottom up – 'ascending' test) are used generally in rock where a good seal can be obtained between the packer and borehole. Rock coring of 76 mm diameter is advised, along with continuous monitoring of the water table. If a packer is set inside the casing to test unconsolidated strata, then care is needed to seal the annulus between casing and hole.

Results can vary from hole to hole in similar strata and are generally more accurate below the water table.

Some relative values of permeability	\multicolumn{10}{c}{Coefficient of permeability k (m/s)}									
	10^{-1}	10^{-2}	10^{-3}	10^{-4}	10^{-5}	10^{-6}	10^{-7}	10^{-8}	10^{-9}	10^{-10}
Permeability	Very high	High	Medium		Low		Very low		Practically impermeable	
Type of soil	GRAVEL		MEDIUM to FINE SAND		V. FINE					
		COARSE SAND		SAND AND SILT						
			FISSURED CLAY					UNFISSURED CLAY		
Type of rock	◄◄ solution features---LIMESTONES---bedding/joints ──►►			Intact ──►►						
		◄◄ coarse---SANDSTONE---fine ──►►								
				SHALE, SILTSTONE						
Fissure interval	<0.2 m	0.2–0.6 m			0.6–2 m			>2 m ──►►		

The coefficient of permeability (*k* in metres per second) is defined in civil engineering as the rate of flow of groundwater through a given area under unit hydraulic head. Strictly, it is the 'hydraulic conductivity' of the soil and should be differentiated from 'intrinsic permeability', which takes account of viscosity. The value of *k* depends mainly on:

- the average porosity, particle size distribution and density of the ground
- the shape and orientation of the particles
- the degree of saturation and air content
- the presence of fissures, stratification and lenses.

Formulae used to calculate permeability depend on the assumption that the stratum being tested is homogeneous. As this is clearly not the case, due to the variations in the soil fabric, no one specific coefficient value can be obtained – only approximations – and care is

Lugeon units are a measure of the potential for rock to absorb water into fissures as assessed by a specific type of *in situ* packer test. One lugeon is defined as water acceptance of 1 l/min through a 1 m length of hole (76 mm diameter) under a head above groundwater level of 100 m for a period of 10 min. The conversion to lugeons assumes that the water flow in the packer test is proportional to the pressure head, which may not always be the case; hence a series of packer tests is made at different pressures. A value of >30 lugeons in a water test indicates high potential absorption; <5 lugeons indicates low.

> One lugeon is approximately equal to a coefficient of permeability of 1×10^{-7} m/s.

Multiple sampling systems provide for groundwater levels, sampling and pumping tests to be carried out in a single borehole at several levels (Soudain, 2003).

Borehole permeameter tests give reasonable estimates in fine soils and provide indications of changes in permeability with change of strata.

Laboratory permeameter tests on small soil samples are unlikely to represent field conditions and should not be relied upon unless correlating data, such as reproducible results on similar samples or other field tests, are available.

Visual examination by an experienced geologist of 'undisturbed' soil samples will assist in determining whether the soil is anisotropic, requiring attention to improve vertical drainage in groundwater control.

Particle size analysis can provide estimates of permeability of soil with less than 20 per cent silt and clay particles, but the method of obtaining the sample will affect the estimate:

- Bulk samples from a shell sampler may not represent the true percentage of fines in the soil – use the minimum estimate obtained from the grading curves from a particular stratum.
- For tests on 'undisturbed' tube samples, the average estimate may be used.

> **Hazen's formula** for approximate values of permeability based on particle size is widely used and gives a reasonable guide for sands with particles 0.1–3.0 mm:
>
> $$k = 0.01\,D_{10}^{2}\ \text{(m/s)},$$
>
> where D_{10} is the 'effective grain size' in millimetres corresponding to 10 per cent passing on the particle size distribution curve.
>
> It is applicable where the 'uniformity coefficient' U ($U = D_{60}/D_{10}$) is approximately 5. Hazen's formula may be applied to soils with $U > 5$ by modifying the 0.01 constant based on experimental data.
>
> **Prugh's method**, using the D_{50} particle size, the uniformity coefficient and the relative density of the soil can give reasonable estimates of permeability over a wider range of particle sizes (datasheets in Preene et al., 2000).

POREWATER PRESSURE

Below the water table the groundwater fully saturates the pores and exerts a 'porewater pressure' (u) which equates to hydrostatic pressure in equilibrium conditions. If the soil is only partially saturated (i.e. the inter-particle space also contains air), then accurate pore pressure measurements can be difficult.

The shear strength of soil and rock (1.3 and 3.1) is related to cohesion, angle of internal friction and porewater pressure. Porewater will reduce both cohesion and friction in soils and weak rocks.

When a load is applied to a saturated soil, the water level in a piezometer with a tip in soil which is undergoing deformation due to the loading will rise above the static water table for a period – this is 'excess porewater pressure' (u_e) and affects ground stability.

The water table represents equilibrium porewater pressure, i.e. at the water table $u = 0$ and at depth z, $u = z\gamma_w$, where γ_w is the density of water.

Monitoring of the water table and porewater pressure should start during the investigation, by installing standpipes or piezometers in strategic positions (where they will not be disturbed during construction) for future control of groundwater systems, dynamic and vibro-compaction of soil, earth embankment construction, etc. Simple observations of water standing in a borehole for a short period are inadequate for most applications.

Porewater pressure will vary with time, climatic conditions and applied load, and it is useful to have ongoing measurements of this parameter before, during and following construction (e.g. in dams and earthworks).

Standpipes are the most basic devices for measuring groundwater levels using a dip metre in boreholes in coarse soils. They comprise open-ended rigid PVC (polyvinyl chloride) pipes 25 mm in diameter with the bottom 1.5 m perforated and surrounded by filter sand over the length of the pipe.

Piezometers use a porous ceramic or perforated plastic element (short-term use only) at the bottom of the standpipe surrounded by filter sand over the response zone. Ideally they do not require porewater to enter the element, but react to changes in pore pressure. Various methods of measuring transmitted pore pressure are used – frequently with data loggers:

- Casagrande piezometers: ceramic or plastic tipped with various means of protection; measurement by dip metre; best suited to saturated soil
- electrical type (vibrating wire): accurate results in most soils; expensive and cannot be used for permeability measurement
- hydraulic type: ceramic tipped; may be used in partially saturated soils (for assessing negative porewater pressure)
- pneumatic type: ceramic tipped; responds rapidly and accurately in saturated soil; not suitable for permeability measurement.

Care is needed in installing these devices either in 150 mm boreholes (particularly when more than one is placed in a single hole, separated by a cement–bentonite grout seal, to check water levels in different strata) or when using drive-in types in soft soil. Time should be allowed for readings to settle down.

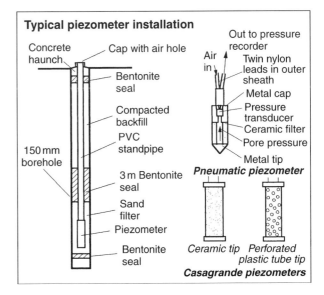

Typical piezometer installation

Concrete haunch — Cap with air hole — Bentonite seal — Compacted backfill — PVC standpipe — 150 mm borehole — 3 m Bentonite seal — Sand filter — Piezometer — Bentonite seal

Out to pressure recorder — Air in — Twin nylon leads in outer sheath — Metal cap — Pressure transducer — Ceramic filter — Pore pressure — Metal tip

Pneumatic piezometer

Ceramic tip — Perforated plastic tube tip

Casagrande piezometers

1.4 Rock Investigations

Generally, intact rock is a strong engineering material. However, because of the history of gravitational, tectonic and weathering forces to which most rock has been subjected (Blyth and de Freitas, 1984), the strength of the 'rock mass' depends on many factors – all of which can affect the design and construction of tunnels and slopes and raise requirements for ground treatment:

- mineralogy
- geological structure and stratification
- density
- anisotropy
- number, state and orientation of 'discontinuities' (i.e. fractures, joints, faults, fissures, shear planes, cleavage, bedding along which movement may take place)
- geomorphology, erosion and weathering
- groundwater and hydrogeology.

Geotechnical processes to deal with the adverse factors resulting from the above are generally costly to install – although, where feasible, relief of groundwater pressure by gravity drainage is the most simple and economic solution for slopes.

Ground investigation should follow the general principles used to prepare the conceptual ground model as the first step. It is essential that standardised descriptions are used for rock cores, discontinuities and weathering (Brown, 1981; Tucker, 1996; Powell, 1998; BS 5930: 1999). In addition the following parameters should be derived to supplement the basic geological cores and laboratory tests to assist in classification of rock mass:

Total core recovery (TCR) is the percentage of the total core recovered compared with the length of the core run from which it came. Minimum specified is usually 80 per cent.

Rock quality designation (RQD) is a measure of the fracture density in a rock core > 50 mm diameter: the percentage of the total length of core pieces > 100 mm compared with the total length of core run. In BS 5930: 1999 75 per cent is classified as 'good'.

Fracture spacing index is the number of fractures per metre length of core.

Weathering grades range from *I* for 'fresh' rock with no visible sign of weathering through to *V* for 'completely weathered' (BS 5930: 1999). Chalk has its own revised weathering grading (1.6).

Rock mass rating (RMR) and the Norwegian Q rating (Hoek and Brown, 1980) provide comprehensive classification of joints, adding scores for different parameters to give guidance on rock slope and tunnel support, e.g.:

- RMR of 80–100 indicates sound rock which can be cut to a near-vertical slope; support is not usually needed in tunnels
- RMR of <20 indicates poor rock; slopes will be <40°; heavy rib support is required in tunnels.

The use of rock mass properties in design generally is examined by Pine and Harrison (2003).

Note that the 'rock mass factor (*j*)' used in BS 8004: 1986 for estimating the deformation modulus (E_m) of rock based on RQD is the RMR expressed as a fraction.

Strength classification ranges from 'very weak (UCS < 1.25 MN/m²)', e.g. weathered chalk, through to 'extremely strong (> 200 MN/m²)', e.g. basalt (BS 5930: 1999).

Orientation of bedding and fractures etc. will determine slope failure pattern and movement in tunnels.

Stereograms using equal area projections (Hoek and Bray, 1981) provide graphical representations of the orientation of the geological structure to aid interpretation and design of slope stability and tunnel support. The stereogram will highlight distinct 'pole' densities around which contours can be drawn to indicate the rock mass structure. A concentration around a single pole will indicate failure due to sliding on a single plane, whereas no pattern to the plotted poles indicates a circular failure. Three-dimensional (3D) analysis of stereograms is now used for applying the New Austrian Tunnelling Method (NATM) in soft ground (Warren and Mortimore, 2003).

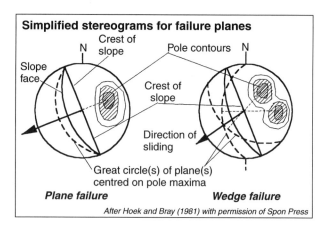

Simplified stereograms for failure planes

Plane failure | Wedge failure

After Hoek and Bray (1981) with permission of Spon Press

Statistical analysis of data by computer is now common, but caution should be exercised as features outside the distribution envelope may be highly significant. (This applies to the selection of all geotechnical design parameters.)

Cross-sections (drawn to natural scale) showing all the relevant geology features and geotechnical data are essential for the design of slopes and tunnels.

UNDERGROUND EXCAVATIONS IN ROCK

Because of the variations in the strength, structure and condition of the rock mass over its length, tunnels and the associated shafts present some of the most difficult investigation, design and construction problems for geotechnical engineers.

In addition to determining the basic geological and soil stratification, permeability and strength parameters, the following have to be considered for structural analysis, alignment and construction:

- major faults, joints and discontinuities and their location and orientation relative to the tunnel or shaft opening
- location and condition of rockhead
- loads from the ground and adjacent structures
- stability of the crown and invert
- excavation methods and spoil disposal
- use of the NATM in soft ground (ICE, 1996b)
- methods controlling groundwater inflow – compressed air and pumping
- presence of toxic and explosive gases, particularly methane (HSE, 1985: Abbeystead explosion report)
- seismic forces
- contingencies for variation in ground conditions.

Geotechnical processes used underground include:

- groundwater exclusion by artificial ground freezing and grouting
- groundwater lowering using external wells
- rock bolting and anchoring
- shotcrete (sprayed concrete).

Ground investigation should be designed in phases to resolve:

- presence of faults, broken ground and buried channels
- volume and rate of inflow of groundwater
- potential for squeezing ground, rock bursts and over-break.

10

The extent and methods of investigation will depend on the material to be excavated. Parameters for tunnelling in London Clay are well known and geotechnical treatment will concentrate on limiting surface movements, whereas the complex geology in parts of the Scottish Highlands will inevitably require more effort to limit unforeseen conditions during tunnelling and to deal with problems which arise.

Desk studies and walk-over surveys (where possible) to produce surface maps are necessary to focus the drilling and geophysical investigations from the surface. Aerial surveys are necessary in difficult terrain. Visible features which may affect tunnel alignment are:

- sinks in the ground profile – indicating possible solution features, old shallow mine workings, seasonal springs, etc.
- landslides and scree at tunnel portals.

Vertical boreholes at 300 m centres are usual, but even at closer spacing unforeseen features (e.g. buried channels, variable rockhead, faulting) can be exposed during construction. Inclined boreholes will be needed where strata, faults or cleavage planes are near-vertical.

Hole depths must be sufficient to ensure that stresses around the tunnel can be assessed and that re-alignment can be made where necessary without additional bore-holes. For most service shafts (as opposed to deep mining shafts) the depth of investigation will not be so great that the borehole wanders off line – provided proper drilling techniques are used.

Blasting efficiency for tunnels and all rock excavation is highly dependent on the rock mass structure and strength. In addition, the *in situ* stresses around the tunnel will affect the amount of over-break, and expert advice should be sought to determine the blast hole pattern and explosive charge required. Safety measures are essential (BS 5607: 1988).

ROCK SLOPES

Some additional geological features which will influence new slope stability and the dimensions of the cutting are:

- steeply dipping beds, producing toppling
- fissures behind slopes, roots in fissures
- fractured rock with low RMR
- weathered rock and overburden over sound rock
- deformation due to *in situ* stresses
- groundwater conditions, producing pressure behind slopes and ice in fissures.

Geotechnical processes used in stabilising rock slopes include:

- rock anchors and bolts
- retaining nets and fences
- dowels
- horizontal drains
- fissure grouting
- surface infilling and shotcrete
- dowelled underpinning.

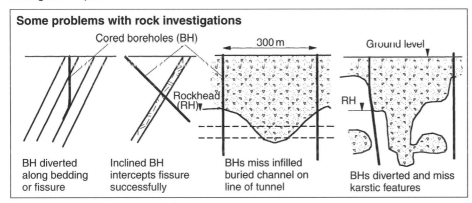

Some problems with rock investigations

BH diverted along bedding or fissure

Inclined BH intercepts fissure successfully

BHs miss infilled buried channel on line of tunnel

BHs diverted and miss karstic features

Ground investigation for the construction of new rock slopes and into the stability of existing slopes requires detailed study of the geomorphology (Fookes, 1997; Fookes *et al.*, 2001) in addition to assessment of the parameters and features listed above.

Boreholes and geophysics should examine rock behind the proposed or existing slope and in front of the toe.

Rippability for excavating rock can be related to seismic velocity (usually *P*-wave) in conjunction with the assessment of the rock mass properties (McDowell *et al.*, 2002). Various types of mechanical breakers are used for cutting slopes adjacent to buildings; the thermic lance will cut most rocks without vibration, but at a high cost.

Phased investigation should continue during tunnelling, particularly geotechnical and geological mapping as rock is exposed.

Groundwater levels and piezometric pressures, particularly in fissures and fractured zones in otherwise intact, relatively impermeable rock, should be carefully measured.

Gases in coal measures are potentially extremely dangerous, toxic and explosive. They can contaminate surrounding strata under pressure and burst into underground excavations. Investigation in such conditions requires specific safety measures (BDA, 2002).

Horizontal probing at the tunnel face in weak rock (as done for the Channel Tunnel in Chalk Marl (Chapman *et al.*, 1992)) ensures adequate notice of groundwater ingress. Effective alleviation facilities have to be readily available on site – drainage, grouting, freezing – and access can be difficult.

Adits may be needed in badly faulted water-bearing ground to assess *in situ* conditions; these may be used for future pre-treatment injection grouting.

Crosshole and downhole geophysics (1.5) provide improved data on discontinuities, isolated voids and strength and stiffness for design of tunnels in hard and soft rock.

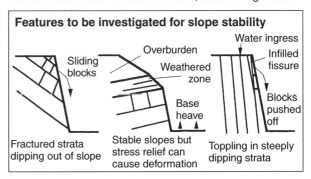

Features to be investigated for slope stability

Fractured strata dipping out of slope

Stable slopes but stress relief can cause deformation

Toppling in steeply dipping strata

Blasting in a steep cutting in most strong rocks will require the use of pre-splitting techniques (i.e. the drilling of a line of closely spaced open holes usually only lightly charged) to limit over-break. It is not suitable adjacent to structures.

1.5 Ancillary Investigations

There are several useful ancillary tools and methods available to the geotechnical engineer to assist in providing economic structures in the ground.

Statutory obligations, risk assessment and health and safety issues must also be considered.

GEOPHYSICAL METHODS

Indirect or non-intrusive methods of obtaining information on the mass of overburden, underlying rock and water table can be useful in enhancing stratification data provided the information is acquired by and interpreted by specialists. The geology over a large site or the status of landfill areas (including the use of airborne instruments (Peart *et al*., 2003)) can be quickly assessed, but such indirect methods should be correlated with borehole data.

Limited geotechnical properties may be obtained – elastic modulus, stiffness, density, porosity, rippability, etc. The improvement of dam foundations by grout injections can be checked by seismic tests (McDowell *et al*., 2002).

The following summarises the main techniques – for which specialist logging personnel are required.

Seismic refraction surveys at the surface require measurement of the velocity through the soil of the primary P (compression) shock wave induced by a hammer blow or small explosion; detection is by a series of geophones. The arrival times of the direct and refracted waves at each geophone allow a profile of seismic velocity versus depth to be plotted. The depth of each layer can then be estimated and interpreted to yield information on depth to rockhead and water table, thickness of overburden and weathering, the presence of faults and buried channels, etc.

Crosshole seismic shots between adjacent boreholes with computer-aided readout can produce data on elastic moduli and identification of major discontinuities. Seismic methods will also give a good indication of rippability and 'diggability' in terms of seismic velocities for specific excavation machines.

Electrical resistivity surveys require measurement of the change in resistance to flow of an applied electric current through different soils, using electrodes driven from the surface – inexpensive. Good indications of rockhead below clay, sands and gravels and some stratification are possible, and detection of sinkholes and shallow cavities in rock is feasible. The method is susceptible to ambiguities, and measurements are affected by high-voltage cables, water-filled cavities, concrete ground cover and buried services.

Crosshole resistivity or tomography can produce a good electrical image of stratification using computer analysis of data from multiple electrode arrays.

Ground-probing radar picks up partial reflections of microwave electromagnetic pulses from shallow (1–3 m) features of bedrock and cavities using a small transmitter and receiver unit pulled over the ground. Rapid detection of buried services and voids near the surface is possible but anomalies can be difficult to interpret.

Gravity surveys measure the variations in the Earth's gravity due to differences in the density of underlying rocks. Mini-gravity surveys are much cheaper and can successfully locate shallow cavities in urban conditions, subject to terrain corrections.

Borehole logging developed from oil exploration technology, can provide useful geotechnical data, particularly on stratigraphy and lithology. BS 7022: 1989 provides information and bibliography on the use of borehole logs for hydrogeological purposes.

PRESSUREMETERS

These instruments are used to determine shear modulus of soil and rock *in situ*, particularly where finite element analysis is used for foundation assessment and design (Clarke, 1996–7).

The shape of the ground response curve obtained from the test depends on the method of installation – pre-bored or self-bored – and requires experience in both installation and interpretation. The engineer must specify the test procedure and method of interpretation in order to obtain the information required (BS 5930: 1999; Eurocode 7 Part 3 (draft), 2000).

Unless the ground is highly anisotropic, the modulus obtained from a pressuremeter test is applicable to most settlement calculations since horizontal and vertical stiffness can be treated as very similar.

In clay it is possible to use the self-boring pressuremeter to estimate *in situ* stress and shear strength (Clarke, 1996).

Pre-bored pressuremeters (Ménard type), 1 m long, 74 mm diameter, are lowered into a cored test pocket in a slightly oversized hole and the probe membrane is expanded to apply a load to the borehole wall. They can be used in most soils and weak rock.

Self-boring pressuremeters replace soil as it is bored; can only penetrate soil and weak rock such as weathered chalk; operate at pressures of around 4.5 MN/m^2.

Push-in pressuremeters, 50 mm diameter, displace soil; pressure capacity half the value of self-boring pressuremeters.

High-pressure dilatometers, 1.5 m long, 73 mm diameter, operate at pressures up to 20 MN/m^2 in preformed slightly oversized holes in weak to moderately strong rocks; the test pocket is 455 mm long. Ideally the test zone should be matched to core samples.

An experienced engineer should evaluate results.

DIRECT OBSERVATION

Observation of quarries and natural faces local to the construction site can provide direct large-scale and detailed information on stratigraphy, lithology, discontinuities, fractures, weathering, etc. which will help in preparing the conceptual ground model.

Photographing of core in colour as it is extracted from the core barrel supplements the logging and description. The following techniques require specialist personnel:

Core orientation devices may be used to supplement the geological markers (bedding planes, discontinuities) to fix the position of the core within its geological context. Most methods mark the first run of core, and care is needed when removing subsequent core and piecing it together for logging.

Borehole camera inspection using closed-circuit television (CCTV) can provide useful direct data on fissure orientation and presence of cavities. The camera can be used in holes >60 mm diameter and can look-forward down the hole and sideways to the walls. This method is also used to determine core orientation.

Acoustic televiewers, which produce images by means of a pulsed ultrasonic beam, work in both mud-filled and dry boreholes. The image can be presented as a solid core with considerable detail. In chalk it is necessary to remove any smearing from the side of the hole before taking the image.

Impression packers press a thermoplastic film on to the sides of a diamond-core-drilled hole in rock using an inflatable rubber packer. This produces an image of the grains of rock and fissures which can be matched to the core and its orientation.

a hole to take BX casing). W denotes the appropriate drill rod/thread.

Most geotechnical drilling produces N and H size cores (54 and 76 mm) – the larger the core, the better is the recovery generally (but at a cost).

Casing of overburden and fractured rock has to be carried out with care to avoid hole collapse and excessive wear on the casing shoe and reaming shell.

Core bits are ring-shaped, with channels for the flushing fluid. The shape of the cutting edge and size of diamond are selected to match the strength and abrasiveness of the rock. Settings range from 10 diamonds per carat for soft rock to 100 diamonds per carat on the face or diamond grit 'impregnated' in a matrix around the head of the bit for strong rock. Manufacturers should be consulted on the type to use. Worn bits must be replaced before the hole and core size are adversely affected.

Core barrels are mainly of the double-tube swivel type mounted on bearings above the bit so that the core and inner barrel are not rotated. The core passes up the inner barrel (which can be lined with a mylar tube to improve recovery in weak rock – the triple system) as rock is penetrated. Core is retained in the tube by a spring assembly (core-catcher) suited to the particular formation. The core barrel is brought to the surface on drill rods after each 'run' (say 3 m of core). Care is required in removing the core from the inner barrel; never hammer the barrel.

Flushing to remove drill cuttings from the hole uses air, water or drilling mud pumped down hollow drill rods (in 'conventional drilling'; see 11.1) to the annulus between the inner and outer barrel, limiting erosion of the core. Discharge is through ports in the face or sides of the core bit.

Rotary core dill

Triplex piston pumps are preferred for water flush, and positive rotary displacement pumps for mud.

Flushing with chemical foams can adversely affect the quality of core, but polymer fluids are now in use.

In situ **tests** can be carried out in stable, unlined cored holes, e.g. high-pressure dilatometer tests, vane tests, borehole CCTV surveys, impression packer tests and core orientation.

Inclined holes can be readily drilled with coring rigs to intersect sub-vertical features such as dykes and faults. It is necessary to survey such holes along their length to check orientation of the core.

Case: Second Severn River Crossing
However detailed the investigation is, only a small portion of the ground is examined. Over 5,000 m of vertical rock core were produced in the investigation for this major bridge from 150 boreholes – this had to cover foundations for a structure 13 km long and constituted < 0.01 per cent of the foundation area.

'Wireline' drilling avoids the loss of time spent in lifting the drill rods and core barrel from deep holes after each core run. Here the core barrel is lowered down the centre of the drill string (parallel wall steel tubes) by wire and latched onto the core bit at the end of the string. The outer barrel and bit are rotated on the string and when the non-rotating inner barrel is full of core, it is winched to the surface. Equipment is designated 'Q', (e.g. HQ), but the core size is about 80 per cent of a coventional core for the same size of hole.

Temporary casing is not used, but if the bit has to be replaced, the whole string has to be removed from the hole, leading to possible collapse.

Flushing volumes and pressures are less than conventional drilling.

Special equipment is available for coring in silts and soft clays.

ROTARY PERCUSSIVE DRILLING

In top hammer drilling the hole is advanced by rotation of the drill rod and bit combined with rapid impact from a hammer in the top-drive motor. Track-mounted air- and hydraulically-powered drills produce rapid, low-cost probe holes when detailed stratigraphy of the rock is not required. They are useful for probing for cavities in rock and drilling overburden using the duplex method of simultaneously drilling/driving 76 mm diameter casing to a maximum depth of 40 m (11.1). The rate of penetration should be recorded.

Flushing media are air, water, mud or foam. Fine cuttings are rapidly brought to the surface with air flush and can be used by an experienced geologist to identify rockhead, major strata changes and the presence of voids – but dust needs to be suppressed and noise abated (HSE, 1997, 1998 and TSO, 2002). Some core drilling to correlate probes is desirable.

Strata changes can only be estimated from changes in colour of the water when using water-flush probing – very difficult to assess in chalk (Lord et al., 2002) and mudrocks. Note must be made of loss of flushing media, indicating fractured and cavernous ground. In water flush the returns may be increased by groundwater flow.

'Down-the-hole' percussive drills are similar to rotary percussive drills, but here the hammer and rotation unit is located above the drill bit at the bottom of the hole. Air flush is used for cuttings inspection; casing is duplex.

There have been useful developments in recent years in power and duplex diameter (11.1). The equipment is more efficient in hard rock.

ROTARY 'OPEN HOLING'

A powerful top-drive rig is used to rotate a 'tricone' rock bit (a full-face toothed wheel bit as developed by the oil industry) through rock – and overburden when casings are adapted. A wide range of depths and diameters is available. The equipment is useful for drilling dewatering wells up to 450 mm, instrumentation holes, grout holes.

Flushing media are air, water, mud or foam. Wash boring is not usually specified for ground investigation in the UK, even in sands and gravels.

CONTINUOUS FLIGHT AUGER (CFA) RIGS

Top-drive rotary drills can be used to rotate augers with a hollow stem into cohesive soil to recover samples accurately and rapidly using a wireline tool. The diameter of the auger flight is usually 250 mm maximum, and depths of 30 m can be achieved to recover 50 mm samples.

No flushing media are required.

2 DECISION-MAKING CHARTS

The selection of an appropriate geotechnical process is based on experience, judgement and cost-effectiveness. The charts in this chapter give some simple guidance on the issues which should be considered initially. Reference can then be made to the information in the text to refine selection.

2.1 Groundwater Control – Removal

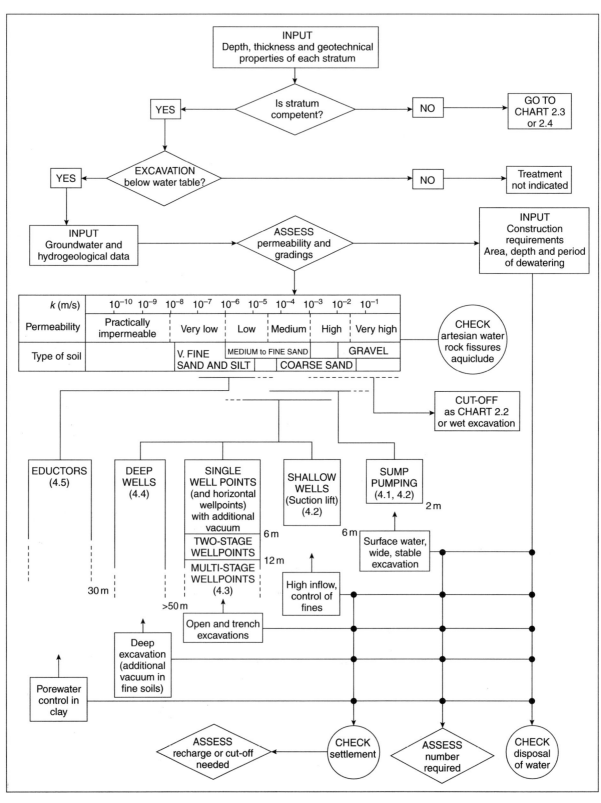

2.2 Groundwater Control – Exclusion

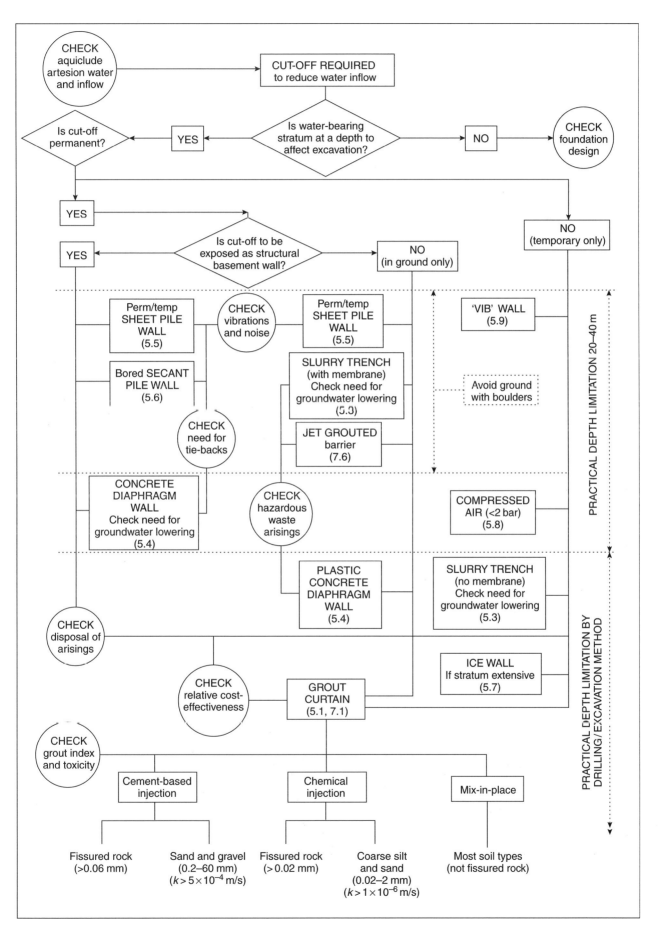

2.3 Ground Improvement 1

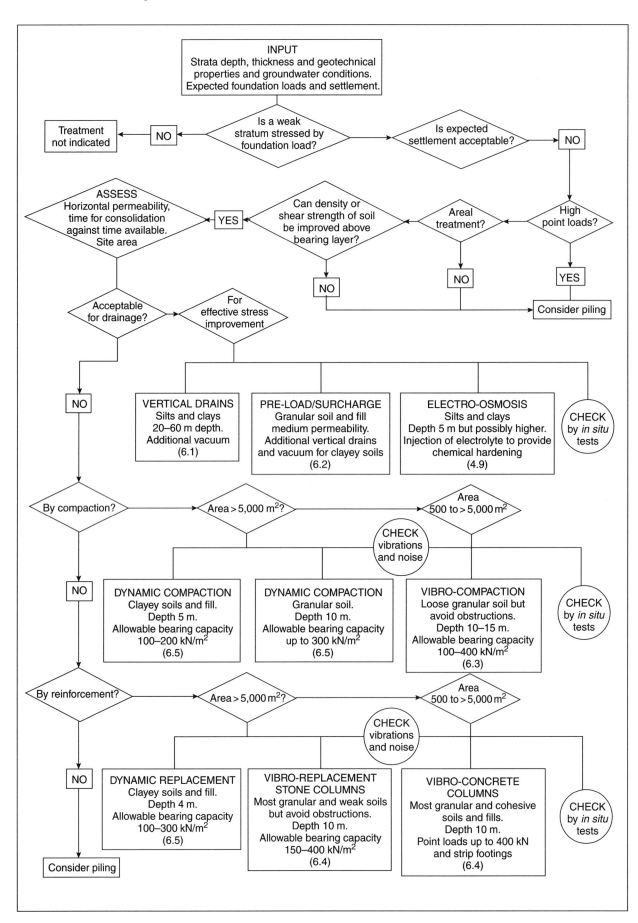

2.4 Ground Improvement 2

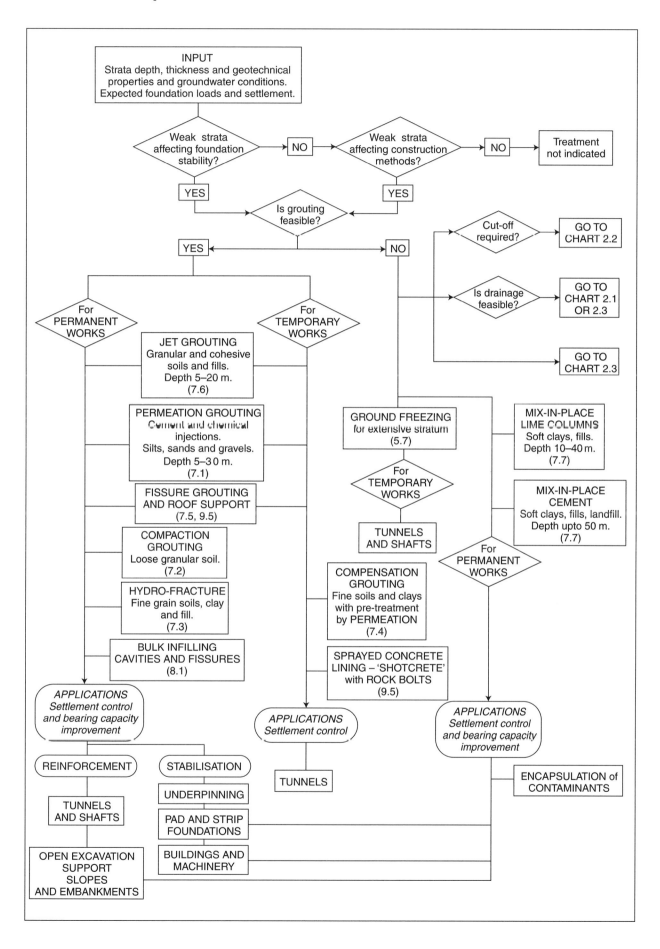

3 GROUNDWATER CONTROL – GENERAL CONSIDERATIONS

3.1 Ground Stability

Groundwater is defined as water contained in pores in soil and fissures in rock below the water table – effectively the surface where the porewater pressure is equal to atmospheric (1.3).

Groundwater is part of the natural hydrological cycle of movement of water as a result of precipitation, percolation and evaporation. The amount of water which can percolate through the ground or run off to the sea, rivers, etc. depends on topography, geology and vegetation. Groundwater levels and flow can cause instability of excavations and foundations by seepage, uplift and lateral pressure. Knowledge of groundwater conditions and soil and rock structure is therefore essential for the safe design and construction of foundations.

Terms used:

- *aquifer* is the water-bearing soil (e.g. sands and gravels) or rock (e.g. limestone and chalk)
- *aquicludes* are soil layers (clays) and intact rock (granite) acting as barriers which limit and confine the gravitational flow of groundwater
- *perched* water occurs when groundwater lies on less permeable lenses of soil embedded in the general aquifer; it may be released into excavations
- *artesian* conditions exist where groundwater is confined under remote hydrostatic pressure in an aquifer by an aquiclude and rises spontaneously above the ground surface when penetrated by a borehole ('sub-artesian' conditions are when the groundwater rises but does not reach the surface)
- *capillary action* may cause porewater to be drawn above the free water surface depending on the soil structure – it is minimal in gravels but potentially in excess of 5 m in silts and fine sands. Surface tension can restrict gravity flow.

Boreholes showing groundwater conditions

After Whitlow (1995) with permission of Pearson Education Limited

Soil structure affects the way groundwater occupies the pores and the performance of the soil as an engineering material when it is compressed under load from buildings or dilated when excavated.

Soil consists of a skeleton of uncemented or weakly bonded mineral particles produced by the weathering of rocks, surrounded by interconnected voids or pores filled with:

- water ('saturated')
- air only ('unsaturated') or
- a mix of air and water ('partially saturated').

The type of base rock and the physical and chemical erosion, weathering and transportation processes which have acted on the rock determine the structure of the soil. For example clays are composed of very fine, plate-like, chemically bonded particles which perform differently from the coarser-grained, rounded or angular particles of silts, sands and gravels, which have little or no bond between them.

Rock structure also affects groundwater conditions and hence performance during construction and long-term stability. Factors include:

- porous mass with cemented particles surrounded by interconnected voids, saturated to unsaturated like soil
- fractures and faults transmitting water under pressure from remote sources
- solution features in chalk and limestone
- infilled fractures limiting or damming flow.

Hydrogeological features influencing stability are:

- virtually full saturation of all soils below the water table, with only a small amount of air in the pores
- artesian flow
- anisotropic permeability and flow
- proximity of adjacent water courses and recharge sources
- seasonal, tidal, barometric and environmental variations in water table and piezometric levels
- flow of groundwater causing 'quick' conditions where upward seepage pressure exceeds the submerged weight of soil.

EFFECTIVE STRESS CONCEPT

When external stress is applied to a soil above the water table the air in the pores will readily compress under load, resulting in the soil structure undergoing a rapid reduction in volume.

Below the water table, the applied stress produces an increase in porewater pressure in the saturated soil. Volume reduction occurs only as water drains out of the pores (water being considered incompressible), at a rate dependent on the permeability and drainage conditions of the soil – rapidly in the case of coarse-grained soils or by slower consolidation in clay soils.

When equilibrium conditions are reached with no flow of water, the total applied stress will, after a time, be shared between the soil particles and the remaining porewater.

Effective normal stress (σ') is defined as the stress transferred through the soil due to the grain-to-grain contact and controls the deformation of a saturated soil.

Porewater pressure (u) acts equally in all directions.

Stress carried by the soil skeleton is therefore the difference between **total normal stress** (σ) and the porewater pressure. Expressed mathematically,

$$\sigma' = \sigma - u.$$

EFFECT ON SOIL STABILITY

Shear strength, defined in 1.2, written in effective stress terms at failure is

$$\tau = c' + \sigma' \tan \phi'.$$

With $\sigma' = \sigma - u$, it is clear that if the porewater pressure can be lowered, the shear strength of the soil and its stability will be increased.

The removal of water from the pores will:

- improve inter-granular contact and density
- cause some deformation (usually settlement) to accommodate the new conditions.

Since capillary action in fine soil is considered as negative porewater pressure, this will also produce an increase in effective stress.

Conversely, stability is reduced when:

- excavation causes gravity flow of groundwater
- excess porewater pressure (u_e) builds up during construction or later loading. Excess porewater pressure will take time to dissipate under load, the process being a function of the soil structure and the history of *in situ* overburden pressure.

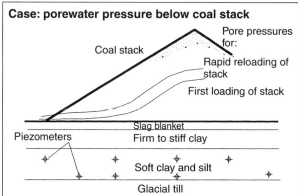

Case: porewater pressure below coal stack

The stacking rate at a large quayside coal-handling facility was 2,000 t/h, capable of building a 17 m high stack in 1 h. At this rate laboratory and field tests showed the soft clay foundation would fail in shear due to the rapid increase in pore pressure. By taking a week to load the clay initially it was shown that subsequent loading and unloading could be done at the design rate without causing failure.
The need for costly sand drains to relieve the pore pressure build-up was avoided.

Draining and consolidation of clay may give rise to immediate problems during excavation and potentially long-term instability, e.g.:

- seepage from fissures and sand partings
- potentially low 'stand-up' time for a cut clay face
- opening of fissures due to stress relief
- time required to reach residual conditions can result in the long-delayed failure of a clay cutting (Skempton, 1977)
- base failure due to exceeding the 'critical depth' (Bjerrum and Eide, 1956).

Similar effective stress processes occur in rock mass, in that as water in fractures is drained, contact of surfaces improves, increasing the resistance to shearing.

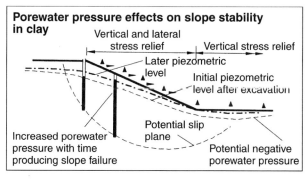

Porewater pressure effects on slope stability in clay

A high water table is one of the most regular and disruptive features encountered in excavations and foundation construction but one which can be effectively dealt with by a specifically designed groundwater control system, properly installed.

The presence of a high water table need not compromise the performance and stability of the foundations – basements, piles, pad footings, etc. – provided that the water table is maintained at the level used in the design or precautions are taken to avoid consequences of raising or lowering the water table.

METHODS OF GROUNDWATER CONTROL

Groundwater control is generally considered under two main applications (see Chapters 4 and 5): groundwater removal and groundwater exclusion.

In addition most sites have to deal with surface water run-off and minor groundwater seepage by gravity drainage and simple pumping away.

Groundwater removal by pumping to allow construction to be carried out safely and efficiently in stable ground conditions by:

- lowering the water table and intercepting seepage to produce dry working conditions
- improving the stability of the sides of excavations
- controlling slope erosion and slumping
- controlling boiling and heave of the excavation base
- consolidating weak strata and improving the shear strength of the soil
- improving groundwater conditions for other geotechnical processes
- reducing lateral loads on earth-retaining structures.

Groundwater exclusion to ensure the stability of the ground and structures by preventing (or limiting to an acceptable degree) ingress of water into excavations and to limit settlement under existing adjacent structures during groundwater lowering using:

- non-load-bearing cut-offs
- load-bearing structural barriers as permanent basement retaining walls.

Combinations of removal and exclusion methods may be appropriate to change groundwater and porewater pressure conditions in the long term to minimise deformations due to effective stress changes.

Temporary groundwater lowering firms up working conditions and provides time and cost savings.

Permanent groundwater lowering can lead to more economic foundation design – particularly important where structures may be affected by rising groundwater (Simpson *et al.*, 1989).

Case: basement in central London
Deep wells have been installed below a new basement in central London ready to be pumped when needed to ensure building serviceability limits and foundation bearing capacity are not compromised by the rising groundwater in the underlying Chalk.

Temporary exclusion by sheet pile walls and cofferdams, artificial ground freezing and grouting not used as permanent works provides dry working conditions within enclosed spaces.

Permanent exclusion by diaphragm walls and bored pile walls used as structural retaining walls around basements, and slurry trenches for cut-offs around landfill sites, also produces short-term construction benefits by providing dry working conditions.

The costs of groundwater control vary considerably, but a carefully designed and installed system can produce economic benefits of improved ground stability, dry working conditions, reduction in weather delays, etc.

3.2 Estimates of Inflow

Seepage or inflow into an excavation is calculated from:

- an empirical/rule-of-thumb basis, experience-based methods
- empirical formulae, either by simple 'hand calculation' or by mathematical modelling
- the observational method, using initial actual flow yields from a chosen system to modify final installations during construction
- flow net methods, either hand drawn or, in the case of complex hydrogeology or large excavations, flow models using computers.

Permeability

All dewatering calculations are subject to the problems of assessing the coefficient of permeability of key individual strata and of the soil or rock as a mass. An apparently modest increase in k from 1×10^{-4} to 3×10^{-4} m/s (within the likely test result range) increases the inflow to a well or excavation by a factor of 3.

The results of the *in situ* and laboratory tests should be interpreted by experienced engineers to ensure that hydrogeological features in the conceptual ground model are considered when assessing this parameter.

Transmissivity

This hydrogeology parameter (T) is the rate at which water will flow through a 1 m wide vertical strip of the aquifer over the full saturated thickness under a hydraulic gradient of 1. ($T = kD$ in metres squared per second for mass permeability.) It governs the ease with which the aquifer will transmit groundwater. Clearly, a change in the water table while dewatering an unconfined aquifer will reduce the saturated thickness and affect the flow.

Specific yield

This is the volume of water (S) which can be drained from an unconfined aquifer per unit area for a unit reduction in head – it is a dimensionless parameter. Water is more easily drained from gravels than from silty sands (S from 0.1 to 0.3). The specific yield is of significance mainly in the initial stages of removing stored water for large excavations as flow calculations are usually based on the 'steady state' following drawdown. Confined aquifers do not exhibit specific yields as such (unless drawdown reduces the saturation level), but a 'storage coefficient' with lower values than the specific yield (<0.01).

Drawdown

This is defined as the degree of lowering of the water table or piezometric level by pumping from an aquifer. In construction dewatering the degree of drawdown will be determined by the requirement for excavation to proceed safely in the dry and be related to the geometry of the excavations. The location of the necessary drains, sumps, wells, etc. will be influenced by surrounding structures, possibly leading to provision of more water extraction points to produce the drawdown than indicated by the volume of total inflow divided by the individual yields. Note that the resulting drawdown may then be greater than required in parts of the site, producing unacceptable settlements unless further precautions are taken.

Zone of influence

This is the radius (R) of the 'cone of depression' in the water table which forms around the drainage element. During drainage the zone of influence will expand until equilibrium (the steady state) is reached. It is an important consideration in groundwater lowering.

METHODS OF CALCULATING INFLOW

Darcy's equation for the laminar (smooth) flow of water through soil is the simplest empirical groundwater formula:

$$Q = kiA,$$

where Q is the flow rate, k the coefficient of permeability, i the hydraulic gradient – the difference in hydraulic head divided by the length of the flow path (dh/l), which conventionally can be a negative term (Freeze and Cherry, 1979) – and A the cross-sectional area through which flow occurs. Increases in k, i and A will produce an increase in flow.

The equation is suitable for estimating total inflow due to vertical and horizontal seepage through uniform soil where flow rates are low.

Dupuit's equation for an unconfined uniform aquifer with circular source can also be applied to near-square excavations to provide a rough estimate of the gravity inflow to the excavation.

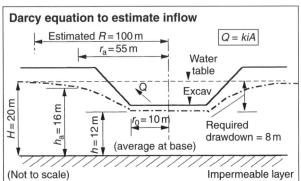

Darcy equation to estimate inflow

$$Q = kiA$$

Estimated $R = 100$ m
$r_a = 55$ m
Water table
Excav
$H = 20$ m
$h_a = 16$ m
$h = 12$ m
$r_0 = 10$ m
(average at base)
Required drawdown = 8 m

(Not to scale)   Impermeable layer

METHOD: (a) Calculate equivalent radius $r_0 = \sqrt{BL/\pi}$ (say, 10 m).
(b) Calculate area A of cylinder at average radius r_a
(= $r_0 + (R - r_0)/2$) where drawdown is h_a, A = $2\pi r_a h_a$ = 5,529 m²
(c) Assess hydraulic gradient i at average radius (= slope of drawdown curve at r_a) = 0.07.
(d) With k = 10^{-3} m/s calculate Q = total inflow = kiA = 0.39 m³/s.

Dupuit well equation to estimate inflow

$$Q = \frac{\pi k(H^2 - h^2)}{\ln\left(\frac{R}{r_0}\right)}$$

METHOD: (Assume conditions as above excavation.)
Substitute drawdown and radius values to give Q = 0.35 m³/s
(= 90% of the Darcy inflow above-reasonably confirmative).

Sichardt's equation is useful to estimate the radius of the zone of influence, R, at steady state of a circular source and radial flow:

$$R = 3000(H - h)\sqrt{k} \text{ (m)},$$

where ($H - h$) is the drawdown (in metres).

It can be applied to excavations if a reasonable approximation to a circle can be made. Another useful empirical relationship for the number of wells, n, required to deal with total inflow, Q, is

$$Q = 3.63 \times 10^{-6} r\, l_w\, n\sqrt{k} \text{ (m}^3\text{/s)},$$

where r is well radius (in metres) and l_w the wetted length of well (in metres).

Jacob's and Thies' equations estimate the coefficients T and S from pumping test data using graphical methods.

Flow nets are graphical representations of the vertical or horizontal seepage flow lines through a soil and the intersecting equipotential lines. They are used to estimate the rate of seepage, porewater pressures and hydraulic gradients in groundwater under sheet piles, under dam cut-offs, in cofferdams, etc. They can also indicate ground instability. *Computer modelling* for flow nets (now including three-dimensional models) has largely displaced the trial-and-error drawing method, but simply sketched flow nets (Cedergren, 1989) will help to ensure that instability due to uncontrolled seepage conditions is avoided in designing a control system.

Instability due to 'piping' (internal erosion in the soil due to flow) inside enclosed cofferdam cut-offs is a function of soil particle size and permeability as well as flow conditions – e.g. piping is unlikely in:

- silty clay, due to its low permeability
- coarse gravel, as large particles may not be displaced by upward flow at the base.

But at the 'critical' hydraulic gradient, the upward seepage force will exceed the weight of soil at the base, reducing the effective stress to zero, resulting in quick conditions and boiling (1.6).

Using a flow net to identify potential quick conditions at the base of a cofferdam excavation, the factor of safety, F (the inverse of the 'exit hydraulic gradient'), must be greater than or equal to 1:

$$F = \frac{N_p a}{H} \geq 1,$$

where N_p is the number of pressure drops from the flow net, a is the width of a net square and H the head of water acting at the base.

If the cofferdam cut-off extends through an aquiclude into an aquifer, the relief of uplift due to piezometric pressure must be considered (4.7). Remedies to stabilise cofferdams other than groundwater lowering include deeper cut-off level (BS 8004: 1986) and construction under water. The following approximation for three-dimensional flow modifying two-dimensional hand-drawn plan flow nets is given by Bowles (1988) for gravity flow conditions into an excavation which fully penetrates the aquifer once steady state has been reached:

$$Q = 0.5k\,(H^2 - h^2)\,\frac{N_f}{N_p},$$

where Q is the inflow, k the permeability, N_f and N_p are the number of flow lines and equipotential drops respectively, H is the depth of the water table and h the depth of water drawn down at the excavation.

The example demonstrates this solution, but caution is needed in applying the method for excavations which only partly penetrate the aquifer due to the upward flow into the base.

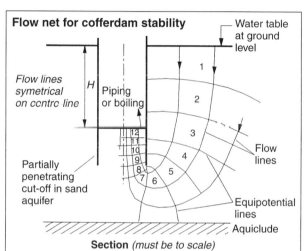

Flow net for cofferdam stability

Section *(must be to scale)*

For this partially penetrating coffer dam, consider $H=12\,m$, and width $=7\,m$ and sustitute in the exit gradient equation. With high water table, $a = 7/7.2 = 0.97$ and $N_p = 12$, $i = 1.03$ and $F = 0.97$, resulting in piping.

If water table is lowered by $1.5\,m$ or more, a will increase slightly, N_p will reduce to 11, to give $F > 1$, eliminating risk of piping.

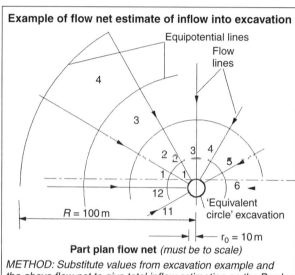

Example of flow net estimate of inflow into excavation

Part plan flow net *(must be to scale)*

METHOD: Substitute values from excavation example and the above flow net to give total inflow estimation as the Bowles equation: $Q = 0.5 \times 1/1000 \times (400-144) \times 12/4 = 0.38\,m^3/s$.

Recharge effects (3.3) will occur when R extends so that the pumping rate is equalled by inflow due to:

- interception of surface water bodies
- vertical recharge from precipitation
- leakage from over- or underlying formations.

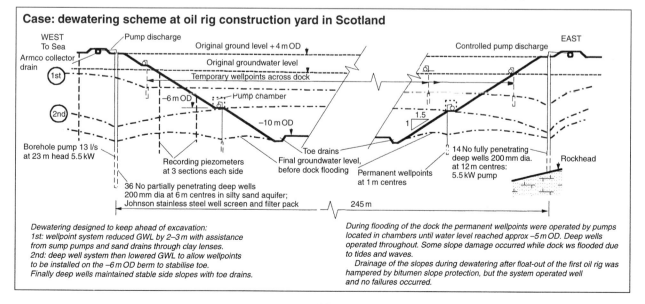

Case: dewatering scheme at oil rig construction yard in Scotland

Dewatering designed to keep ahead of excavation:
1st: wellpoint system reduced GWL by 2–3 m with assistance from sump pumps and sand drains through clay lenses.
2nd: deep well system then lowered GWL to allow wellpoints to be installed on the –6 m OD berm to stabilise toe.
Finally deep wells maintained stable side slopes with toe drains.

During flooding of the dock the permanent wellpoints were operated by pumps located in chambers until water level reached approx –5 m OD. Deep wells operated throughout. Some slope damage occurred while dock ws flooded due to tides and waves.
Drainage of the slopes during dewatering after float-out of the first oil rig was hampered by bitumen slope protection, but the system operated well and no failures occurred.

3.3 Design of Groundwater Lowering Systems

Ground investigations as described in Chapter 1, followed by the general assessments as noted in 3.1 and 3.2 above, should be undertaken before any design for groundwater lowering for construction (dewatering) is contemplated; additionally, engineering judgements have to be applied to ensure safe, practical and economic design.

The conceptual model, including hydrogeological and soil mechanics information developed from the ground investigation and the construction requirements noted above, should be re-visited to ensure relevant data are included, specifically:

- range of permeabilities (check for anisotropy)
- depth of aquifer
- confining and perched aquifers, aquicludes
- artesian or sub-artesian conditions
- boundaries of relevant aquifers
- potential recharge from nearby rivers or fissured rock
- potential lowering of remote surface water levels
- seasonal and tidal variations in the water table
- thin permeable layers and partings in the soil matrix.

Construction considerations:

- drawdown must be at least 0.5 m below excavation depth
- the zone of influence of dewatered ground affecting other structures
- adequate wetted length of the well at steady state
- recharge of the aquifer under adjacent structures to avoid settlement
- cut-off barrier to avoid removing groundwater from under adjacent structures
- extra excavation to accommodate drainage blankets to control removal of fines
- time available to achieve drawdown
- period of dewatering
- disposal of extracted groundwater in accordance with environmental constraints.

Environmental considerations:

- check the need for licences from the Environment Agency (or statutory authorities in Wales and Scotland) under the various Environment Acts and locally delegated powers to 'abstract' or 'impound' groundwater, or for a 'discharge consent' (Masters-Williams et al., 2001)
- discharge of drilling fluids and solids
- consent from water company to discharge groundwater into local stormwater sewers
- effect of drawdown on local water supply wells or other adjacent groundwater control systems
- potential for migration of contaminants from adjacent brownfield sites
- intrusion of saline water into an aquifer near the sea
- contract-specific requirements on pollution control.

Design calculations

The basic differences between designing a well for dewatering an excavation and designing one for water supply are:

- the temporary nature of site work
- the need for multiple wells
- rapid drawdown of the water table to suit construction.

Steady-state laminar flow calculations appropriate for water supply in a uniform aquifer will apply to dewatering only when the required drawdown has been reached following initial maximising of water extraction – possibly under non-laminar ('turbulent') flow conditions.

The designer should aim to provide a flexible dewatering system which can be readily modified to meet unforeseen conditions. This will require an iterative approach to design using several trial parameters to produce schemes which can be checked for cost-effectiveness.

Reliable pumping tests provide the most appropriate design data. If the data are obtained at steady state and laminar flow, and given the dimensions of the aquifer, estimates can then be made of permeability, transmissivity, storage and long-term drawdown using empirical and theoretical well hydraulics.

Non-equilibrium tests are quicker to perform and, provided that time versus drawdown and water table recovery data are obtained, are acceptable for estimating aquifer performance.

Where pumping tests are not available, and particularly in confined, non-uniform aquifers, more reliance has to be placed on experience and judgement to estimate both the total inflow into an excavation and the yield of individual wells.

Yield or quantity of water (q) to be pumped by individual elements in the dewatering system to cope with the total inflow (Q) to an excavation can be estimated from the Darcy equation (which assumes a circular source and radial flow to the well). Yield calculations at steady flow in more complex situations (fully/partially penetrating well, circular/line source, etc.), using the methods of Dupuit, Thies, Jacob and others, are given in Cashman and Preene (2001) and Preene et al. (2000).

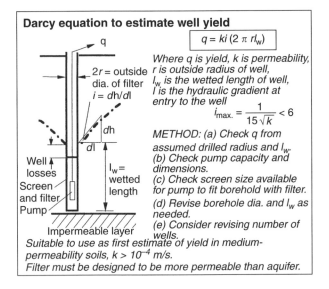

Darcy equation to estimate well yield

$$q = ki\,(2\,\pi\,rl_w)$$

Where q is yield, k is permeability, r is outside radius of well, l_w is the wetted length of well, l is the hydraulic gradient at entry to the well

$$i_{max.} = \frac{1}{15\sqrt{k}} < 6$$

METHOD: (a) Check q from assumed drilled radius and l_w. (b) Check pump capacity and dimensions. (c) Check screen size available for pump to fit borehole with filter. (d) Revise borehole dia. and l_w as needed. (e) Consider revising number of wells.

Suitable to use as first estimate of yield in medium-permeability soils, $k > 10^{-4}$ m/s.

Filter must be designed to be more permeable than aquifer.

(Diagram labels: q; $2r$ = outside dia. of filter; $i = dh/dl$; dh; dl; Well losses; Screen and filter; Pump; l_w = wetted length; Impermeable layer)

The number of dewatering wells (or wellpoints) can be calculated from the total inflow and the individual well yield or the empirical relationship in 3.2.

Spacing of wells is a function of the particle size of the aquifer and permeability: closer in fine soil, further apart in coarse. But for most sites, in order to accommodate the number estimated, the locations will be dictated by the excavation geometry, speed of drawdown and the need to ensure overlap of drawdown in the event of a pump failure.

Graphical design methods, using distance versus drawdown when pumping test results are available (as in the simple example below), give a quick assessment of yield and drawdown for dewatering using wells and are useful when applying the observational method to a series of wells as each well is brought on stream.

The distance versus drawdown line (as in the stage 1 diagram) has a slope which is twice that of the time versus drawdown line for the same pumping rate; hence additional distance versus drawdown lines can be constructed for different times. By constructing distance versus drawdown lines from derived time versus drawdown lines for different pumping rates, a distance versus drawdown line can be drawn for any pumping rate and period of elapsed time.

Design using pumping tests – stage 1

Distance v specific drawdown graph

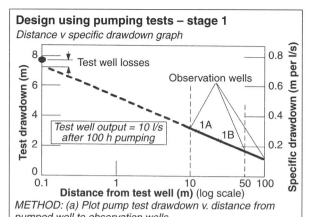

METHOD: (a) Plot pump test drawdown v. distance from pumped well to observation wells.
(b) Draw specific drawdown graph by dividing test drawdown by well output at end of tests (right-hand scale above).

Design using pumping tests – stage 2

Required drawdown in excavation

Plan to excavation

Section XX

Design using pumping tests – stage 3
Calculation of yield required to produce drawdown

Well No.	Distance to point A (m)	Specific drawdown (m per l/s)	Distance to point B (m)	Specific drawdown (m per l/s)
1	12.0	0.32	22.0	0.26
2	15.0	0.29	24.2	0.24
Cumulative		0.61		0.50
Drawdown required	4 m		2 m	
Well yield required	4/0.61 = 6.6 l/s		2/0.5 = 4.0 l/s	

METHOD: (a) Design each well to yield 6.6/90% = 7.3 l/s.
(b) This yield at well 1 will produce drawdown of 2.4 m at well 2.
(c) Pumping at well 2 only at this yield will produce drawdown at well 2 of around 5 m.
(d) Hence drawdown at well 1 and 2 will be around 7.5 m to produce the required 4 m drawdown at point A.

A time versus drawdown graph can indicate recharge or limitations of the aquifer. In addition, as drawdown is proportional to pumping rate, drawdown can be estimated at any time under different pumping rates, and time versus drawdown lines can be drawn for different rates – giving projections of future drawdown at variable rates.
Stored water volume to be removed (i.e. from water table level to equilibrium drawdown level) can be calculated from transmissivity and specific yield estimates to indicate additional pumping and time required.

Time v drawdown graph

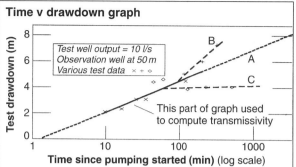

Line A extropolation gives drawdown at 50 m from well after 1,000+ min of pumping.
Increase in slope in alternative line B indicates a limited aquifer (also indicates interference from other wells in a multi-well system as zones of influence intersect).
Reduction in slope in alternative line C indicates recharge of aquifer within zone of influence.
Other useful data determined from this graph are transmissivity and the storage coefficient.

Settlement as a result of reduced porewater pressure surrounding the dewatered excavation is unlikely to be significant in soils which can be dewatered. Where compressible layers of finer soils and peat exist within and above the aquifer, then settlement can occur over a prolonged period (1.2). Settlement due to removal of fines during pumping is avoided by properly designed filters around the dewatering elements. Risk assessment of potential damage to adjacent structures should be carried out as part of the design.

Case: a tunnel being excavated below the water table in limestone drained groundwater not only in the limestone but also from overlying silty clay. The resulting desiccation and shrinkage caused settlements of 100 mm to town buildings.

Pump selection for all groundwater removal will require the assessment of:

- quantity required, nature of service and environment
- capacity of the system and each pump
- static suction and discharge heads
- friction losses from pipes and fittings
- application of additional vacuum to pumps
- discharge to the atmosphere or pressurised main
- nature of the water and likely solids content
- critical time to ground failure in the event of pump failure.

Selection will also be influenced by the plant which is available and the site locations for installation – in wells, wellpoints or sumps – near to buildings, underground, in or clear of excavations.
Chapter 4 provides some typical specifications of pumps suitable for sumps, wellpoints and wells. Data on friction losses in pipes and pipe-fittings are given in Preene *et al.* (2000).

Caution: adequate standby capacity should be provided for each element in the groundwater control system (power, pumps and monitors) and be capable of being brought into rapid service in the event of a failure in any element, particularly during the initial drawdown period. Thereafter it may not be economical to provide full standby plant, which is acceptable provided that the adjacent pumps/system can cope with a short-term failure and replacement is achieved within the critical time.

3.4 Design of Groundwater Exclusion Systems

Groundwater control for excavations can be achieved by exclusion or partial exclusion in conditions where removal is not feasible or economic. Barriers to the flow of groundwater may be required:

- as part of the permanent works
- to protect buildings adjacent to excavations
- to reduce flow to a dewatering scheme for a deep excavation adjacent to the sea or river
- to reduce rate of flow through a ground-freezing zone
- to reduce uplift pressure
- to reduce the area required for side slopes to excavations
- to control migration of pollutants and gases.

Vertical barriers (or 'cut-offs') forming the exclusion system should ideally extend into an impermeable stratum, within an economic depth, to avoid upward seepage into the excavation during construction. If the cut-off is not sealed into an impermeable layer for permanent works such as a basement, then the basement must be designed to resist uplift for its serviceable life.

Horizontal barriers may be necessary below the excavation between the vertical cut-offs where an impermeable stratum is not available to control seepage.

Selection of the exclusion method therefore depends on the purpose, whether temporary for the construction period only or as a permanent structure, the period available for construction, the ground conditions and costs.

The conceptual model developed from the ground investigation should provide relevant geotechnical data for selection of the method and design parameters for these structures (AGS, 1998). For all the barrier methods considered, information will be required on:

- stratification and lithology
- soil density and in-ground obstructions such as boulders and manmade debris to avoid:
  - clutch separation and toe distortion when driving steel sheet piles
  - collapse of narrow slurry trenches when removing large boulders from side walls
  - diversion of drill holes for grouting and ground-freezing tubes leading to untreated 'windows'
- depth to and thickness of the impermeable layer for barrier toe-in
- depth to the bearing layer for piled cut-offs and structural diaphragm walls
- permeability, particle size analysis and fissuring of individual strata for injection grouting applications
- chemistry of soil and groundwater for grouting and slurry applications
- underground services.

Design may be grouped into exposed and in-ground barriers.

Exposed barriers, whether temporary or permanent, are designed as vertical, embedded earth-retaining structures, e.g. basement walls formed by:

- sheet piles
- diaphragm walls or
- secant/contiguous bored pile walls,

using the accepted and codified retaining wall principles (BS 8002: 1994; ICE, 1996a; Puller, 1996; Eurocode 7 (draft), 2000; Gaba *et al.*, 2003).

Design calculations to resist bending and displacement of the structural elements are required at various limit states to ensure equilibrium conditions under relevant lateral earth pressures.

Information on soil strength and stiffness, the angle of shearing resistance at peak residual and critical strengths, particle size distribution, consistency limits and comprehensive groundwater levels is essential. Drained and undrained conditions may be needed as pore pressures may vary after construction.

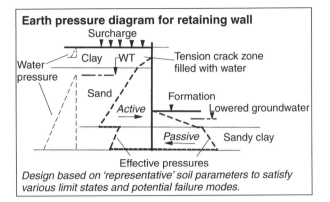

Earth pressure diagram for retaining wall

Design based on 'representative' soil parameters to satisfy various limit states and potential failure modes.

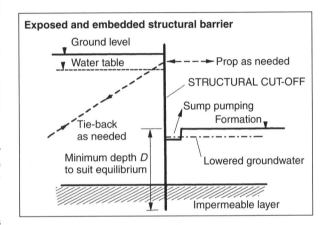

Exposed and embedded structural barrier

Permeability of the permanent retaining wall itself (steel or reinforced concrete) is usually very low, and the joints of a well-constructed basement wall should not allow significant seepage – avoiding additional costly drainage measures during both construction and service life.

Temporary, exposed barrier walls may not have the same attention paid to durability and sealing of joints, leading to additional pumping during construction.

For large excavations enclosed by cut-off walls but with the potential for seepage at the base, supplementary pumping can be assessed using the pumping tests devised by Monnet and Iagolnitzer (1994).

In-ground barriers (cut-offs constructed below ground level) are also used for temporary and permanent applications, but do not have an exposed excavation face and have little or no capability to resist lateral earth pressure (except possibly as a coherent gravity block):

- slurry trenches – a recent application includes reinforcing with bored piles to act as temporary retaining walls
- plastic diaphragm walls
- grout curtains created by jet grouting and by permeation (which in the right conditions can retain soil)
- artificially frozen ground, which may be exposed during excavation for shafts and in certain conditions as propped retaining walls.

Vertical cut-offs are used to prevent groundwater flow into excavations and seepage and pressure head under hydraulic structures such as dams, weirs and reservoirs.

Horizontal and inclined cut-offs can be constructed using permeation and jet grouting methods. It is unlikely that a horizontal barrier could be constructed economically to prevent or even marginally reduce leakage from the base of a large reservoir on fissured rock (unless investigations could identify specific zones).

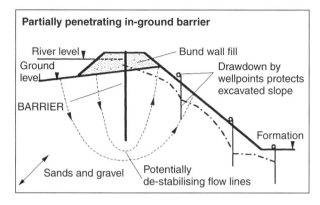

Partially penetrating in-ground barrier

Contaminated land containment and control of groundwater leachate using in-ground barriers is covered in detail in Privett *et al.* (1996), together with associated surface containment barriers. In these circumstances a specialist geoenvironmental investigation and full risk assessment of the source–pathway–receptor route for potential harm must be carried out (BS 10175: 2001).

Design calculations for long-term use are based on factors such as:

- flexural strength
- minimising strain and permeability
- resistance to the hydraulic gradient across the width of the barrier to avoid erosion/solution
- durability and defects control (McQuade and Needham, 1999)
- reduced permeability using a central membrane.

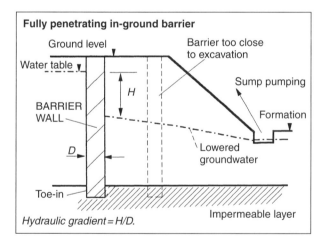

Fully penetrating in-ground barrier

Hydraulic gradient $= H/D$.

Permeability of plastic concrete diaphragm wall cut-offs will be low, subject to good panel joints. Permeability of alluvial grout curtains at best will be 10^{-7} m/s; of slurry trenches, 10^{-8} m/s, but with a central membrane this can be reduced to 10^{-10} m/s.

Durability of in-ground barriers is not well enough researched to show how permanent is a 'permanent' cut-off. Grout curtains in rock under dams have been effective for 100 years, and plastic concrete and jet grout walls properly constructed should have similar longevity. Slurry trenches and chemical grout curtains (Caron, 1963, on silicates) may be more vulnerable to distortion and erosion, reducing design life to, say, 25 years – which may be adequate for industrial developments.

Construction details will depend on whether an exposed or in-ground barrier is selected, but the following features

will always have to be considered by the designer:

- site access and area available for construction
- availability of materials
- environmental issues – noise, vibration, dust controls, disposal of used slurries
- settlement of adjacent structures
- backup if barrier is breached
- maintenance – corrosion of steel, erosion of slurry.

Supplementary pumping during construction will be required to remove 'stored' groundwater from the soil and rock enclosed by the exclusion barrier; also possibly for pressure relief.

Materials for in-ground barriers are covered in Chapter 5 for each process.

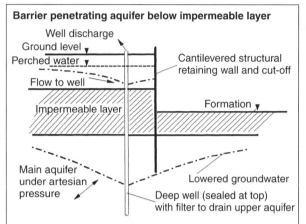

Barrier penetrating aquifer below impermeable layer

Uplift pressure on base of excavation relieved by pumping from lower aquifer. With filter extending up the well some drawdown will be achieved in the upper perched aquifer, reducing lateral pressure on the cut-off wall.

Grouted barriers can be formed as an annulus or 'aureole' around and ahead of tunnels and shafts to reduce water inflow using permeation and fissure grouting techniques. Ice walls serve a similar function.

'Grouted blanket' barriers can be required below dams in addition to the vertical barrier and downstream drainage to control erosion where there is excessive groundwater flow.

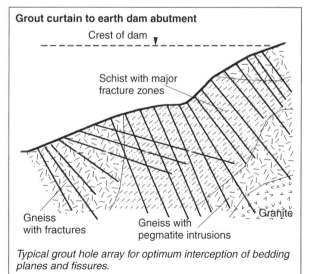

Grout curtain to earth dam abutment

Typical grout hole array for optimum interception of bedding planes and fissures.

Compressed air is used to exclude water in caisson sinking for bridge piers (Tomlinson, 2001) and in tunnels. This is a complex construction method and outside the scope of this *Introduction*, but some general geotechnical points are covered in 5.8.

3.5 Monitoring of Groundwater Control

Monitoring of the performance of the dewatering system or groundwater barrier is essential to ensure that the targets set have been achieved and work in excavations can be carried out safely.

The soil profile should be sampled and logged as each well and piezometer is drilled and checked against ground investigation data. Logging of wellpoint jetting or slurry trenches during construction will not be particularly accurate.

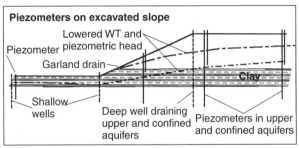

Piezometers on excavated slope

Water levels should be checked and recorded regularly to show drawdown achieved at the well and in strategically placed piezometers or standpipes and in un-pumped wells. If there is a short time to ground failure due to pump failure then frequent readings (every 3 h, say) are required.

Dip metres are the most common means of taking water levels in wells and standpipes, but care is needed for reliable readings: the well may be clogged, or moisture may be picked up higher up the well. Recording piezometers (data-loggers) with direct downloading of data to the responsible engineer are useful for long-term and critical applications.

Water levels in wells should be controlled with high- and low-level electrodes so that the pump is protected and water is maintained at the designed level. Head losses through the filter and screen and turbulence from the pump may mean higher water levels exist just outside the well.

Note that the movement of groundwater is a three-dimensional problem and requires expert assessment.

Discharge rate should be measured and recorded regularly using flow metres, weir tanks, venturi flumes or simple volume/time checks. Measurements will be needed for individual wells and, on large schemes, the collected total discharge. Wellpoint and eductor discharges will be measured at a collection point.

Wells and Wellpoints around temporary dry drock

The fines content of the groundwater extracted from sites must be monitored and action taken to prevent silt and sand entering natural water courses. The Environment Protection Acts provide for penalties against those allowing silt to be deposited in rivers.

Soakaways for discharge may avoid the need for settling tanks to remove fines, but are impractical if discharged water re-enters the zone of influence of the dewatering or if they silt up.

The quality of the discharge water in industrial areas and adjacent to landfill sites should be checked for potential contaminants which may need onsite treatment before discharge or may have to be carted off site as required under the Control of Pollution Act (discharge consent).

For long-term pumping, water quality checks are needed for assessing potential clogging of the well screen by chemical encrustation, corrosion and bacterial growth (Powrie *et al.*, 1990).

Where in-ground barriers are used to contain pollutants, the measurement of water quality downstream of the barrier will have to provide for collecting samples to check attenuation of the different contaminants at varying distances from the barrier. This will require specialist suction and displacement samplers at several levels in the monitoring borehole – within high-density polyethylene (HDPE) casing and screen to avoid contamination by trace elements.

Settlement of structures identified as being at risk should be checked by regular level and crack surveys and compared with the pre-contract condition survey.

Hydraulic gradient efficiency across an in-ground barrier should be checked with piezometers inside and outside the enclosed area.

Plant and equipment performance is the key to effective dewatering, and regular checks on power supply and cables, fuel supply and the preventive maintenance schedule for pumps, generators and pipework should be provided through regular site personnel attendance. Visible and audible alarms to indicate failures should be installed at critical locations.

Maintenance checks may also be needed on the well itself where encrustation and corrosion are likely in long-term pumping.

Tophead-drive rotary rig drilling dewatering well

Records of all actions and observations taken are essential to help with diagnosis of any technical or contractual problems which may arise.

3.6 Impact of Groundwater on Other Geotechnical Processes

The presence of groundwater can influence the effectiveness of specialist construction techniques and geotechnical processes.

Diaphragm walls

Dewatering may be needed to undertake slurry trench and diaphragm wall construction (cases in 5.4) so that adequate head of the support slurry is available above groundwater to prevent slumping.

Diaphragm wall joints can be made waterproof to a considerable degree by casting in a water-bar.

Secant bored pile walls, fully interlocked to act as cut-offs, are likely to have low-permeability joints,

Shotcreting to bored pile wall

whereas contiguous bored pile walls may need grouting or shotcrete at the joints in order to control water inflow.

Bored piles

The stability of bored pile holes can be improved by dewatering during construction; this can also improve the shaft friction between the pile and the re-saturated ground as a result of the improved quality of concrete placed in the dry.

Ground freezing

Groundwater flow or tidal influence on groundwater may need to be reduced for artificial ground freezing to be successful. The ice wall may be distorted by different rates of flow of groundwater and different thermal conductivities in different layers of soil. On the other hand, there must be adequate groundwater present to form an intact ice wall.

Capillary attraction in fine soils may increase the water content as the soil is frozen. This may lead to much reduced effective stress and settlement on thawing.

Extensive freezing will take a long time to thaw.

Alluvial grouting

In alluvial grouting it is necessary to avoid trapping porewater between adjacent grouted 'cylinders' which is difficult to displace with secondary and tertiary injections of grout. Alluvial grout curtains provide only a modest reduction in permeability, and spending additional time on expensive chemical injections is rarely cost-effective. High groundwater flow will cause displacement of grout away from the desired permeation zone.

Slurry trenches

Where these are used as contaminant and methane cut-offs, gas permeability is reduced with a central HDPE membrane.

Contaminated groundwater will affect the durability and stability of the slurry.

Jet grouting

Provided that the cut-off can be formed with interlocking grouted cylinders, the joints will be relatively impermeable. This is feasible in moderate groundwater flow conditions. Mix-in-place methods can be successful in soils with high moisture content (>150 per cent).

Contaminated land

Considerable benefits can be achieved for redevelopment if the groundwater quality can be improved by abstraction using wells and treatment prior to discharge – with due note of environmental constraints.

Effective stress

Groundwater lowering will reduce porewater pressure and increase vertical effective stress around the excavation during construction; when dewatering is stopped, assuming that total stress remains the same, effective stress could return to the original conditions. The design of foundations must therefore provide for both drained and undrained conditions in the ground and accommodate anticipated settlement or heave.

The changes in effective stress may lead to differential settlement which may cause structural damage over a wide area (Preene, 2000). The assessment of the risks of damage is a complex three-dimensional problem solved by numerical modelling for large sensitive projects. For smaller sites such analysis is usually not warranted, but a pre-construction survey of buildings within the zone of influence is recommended. Settlements can also occur as a result of removal of fines during uncontrolled dewatering for excavations.

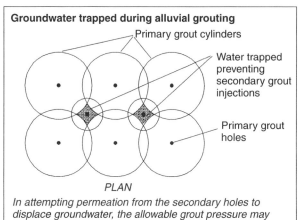

Groundwater trapped during alluvial grouting

In attempting permeation from the secondary holes to displace groundwater, the allowable grout pressure may be exceeded resulting in rupture of the ground and surface heave.

Hydraulic gradient

This is important where internal erosion may be caused within slurry trenches, alluvial grout curtains and thin membrane walls if the differential head produces groundwater flow.

4 GROUNDWATER CONTROL – REMOVAL METHODS

4.1 Gravity Drainage

Groundwater will flow under gravity towards drains, ditches, wells, etc. in soils with permeability $k > 5 \times 10^{-5}$ m/s. In fine-grained, less permeable soils, the gravity flow is much reduced, and measures to deal with pore-water pressure and improve flow (e.g. vacuum wells and pre-loading) are required to ensure stability of excavations. Provided the drainage path is controlled, simple gravity drainage is a cost-effective measure which does not always require specialist dewatering design.

Suitable for controlling surface water run-off and groundwater in sands and gravel by open ditches, 'French drains' (perforated or porous pipes in trenches surrounded by filter gravel), 'grips' (open drains in excavations), soakaways and garland drains on slopes.
Gravity drainage of groundwater into ditches and drains is not particularly effective for slope stability in fine soils – slumping is a common feature.

Design and installation largely depend on site layout, levels and location. The main requirement will be to construct drains and trenches to levels and falls to intercept surface and groundwater flow paths, which avoid ponding of water and are in locations which do not interfere with site traffic and permanent works. Some examples follow:
Surface drains should be designed to collect run-off before it reaches the area immediately behind the crest of a slope, in soil or rock, where the potential for tension cracks exists. It may be useful to line these drains with heavy-duty plastic to prevent water entering cracks and causing slope instability by surface seepage or internal pressure. Ponding of water behind the crest must also be avoided for the same reasons.
Grips at the toe of excavated slopes will help stabilise rock slopes and slopes in coarse granular soils. Also drainage of the excavation formation will be aided. Pumping away is usually necessary.
French (or trench) drains with porous pipes surrounded by graded filter are very versatile for controlling flow of fines from the soil in shallow and deep drains. They are used for slope stabilisation (Bromhead, 1992), motorway and rail track drainage and general site drainage.
Herringbone drains on the face of a slope are effective both in catching run-off and in draining groundwater from the slope. Collector drains are required at the toe to ensure water does not re-enter the slope.
Garland drains may be used to intercept seepage from a permeable layer overlying a less permeable stratum or rockhead. The collected water may be drained through vertical drains to a lower-level water table (which may have to be pumped away).

Surface protection against erosion due to run-off can be provided by laying plastic sheeting and by asphalt lining on soil slopes and by gunite or shotcrete on rock slopes. Weep-holes will be needed to avoid water pressure building up below these linings. Drainage blankets, as noted under sump drainage (4.2), also provide surface protection of soil slopes.
Drainage galleries and shafts from which arrays of drains are drilled can effectively drain slopes, particularly if the gallery is placed near the toe (Hutchinson, 1984) and water can drain under gravity from the gallery.

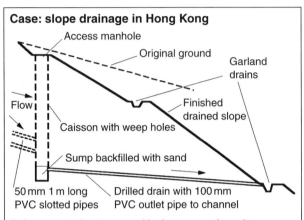

Case: slope drainage in Hong Kong

A slope was to be steepened in decomposed granite. Groundwater was drained into 2 m diameter hand-dug caissons through weep holes and short lengths of drain. The water collected in the sump was run off through a drain drilled from the sump to the lined surface collector channel

After McNicholl et al. (1986) with permission

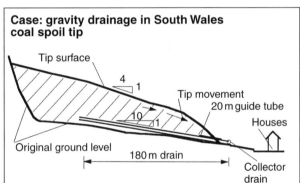

Case: gravity drainage in South Wales coal spoil tip

The spoil tip was moving dangerously close to housing. 10 No 50 mm perforated PVC drains were inserted inside 150 mm temporary casings drilled with directional control into the spoil. Movement of the tip ceased within 6 months of installation. The drain can be cleaned by back-flushing as needed.

Near-horizontal drains can be drilled for long distances into slopes – from the toe preferably – to intercept and drain groundwater without the need for pumping. They are useful for draining water in tension cracks, failure surfaces and acting as general weep-holes in rock.
Vertical sand drains and pressure relief wells are also forms of gravity drainage (4.7, 6.1).

Caution: all drainage works will require some form of control at the outfall to deal with any suspended solids before the water can be discharged into local water courses or sewers. Regular maintenance of the drainage system will help avoid siltation.

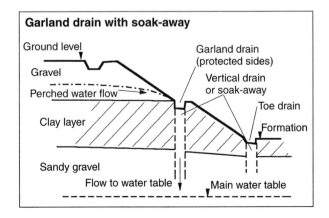

Garland drain with soak-away

Installing submersible pump in small diameter well

Typical submersible sump pump

4.2 Sump Pumping and Shallow Wells

SUMPS

Sumps are a simple and generally cheap method to apply, given proper precautions. They are usually undertaken as temporary control by the civil engineering contractor to collect water from gravity drainage to toe drains, French drains, garland drains, grips, etc. (4.1).

Suitable for control of groundwater in open, shallow excavations in clean gravels and coarse sands ($k > 10^{-2}$ m/s); open fissures in rock:

- Can deal with any quantity of water by using the appropriate number of suitable pumps, but sumps must be carefully located and supported in cases where high inflow could cause erosion of the excavation side slope.
- Can encourage slumping of excavation if fines are removed or hydraulic gradient is excessive.

Not recommended in uniform fine sands, silts and soft clays, soft rock, and particularly where excavated side slopes do not readily drain and have to be flattened for stability.

Design of sumps for open excavations – number and location – is essentially based on judgement, provided that the soil is easily drained. The sump should be deep enough to drain the excavation and ensure that the sump pump intake is submerged.

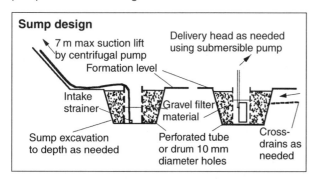

Sump design

7 m max suction lift by centrifugal pump
Delivery head as needed using submersible pump
Formation level
Intake strainer
Gravel filter material
Sump excavation to depth as needed
Perforated tube or drum 10 mm diameter holes
Cross-drains as needed

Where there is a risk of seepage emerging on to a slope, a filter blanket to suit the soil grading can be effective in maintaining stability, particularly when used with a geomembrane or a herringbone pattern of slope drains.

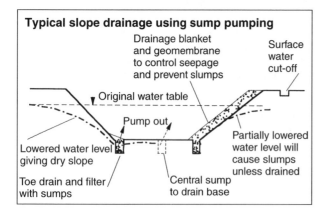

Typical slope drainage using sump pumping

Drainage blanket and geomembrane to control seepage and prevent slumps
Surface water cut-off
Original water table
Pump out
Lowered water level giving dry slope
Toe drain and filter with sumps
Partially lowered water level will cause slumps unless drained
Central sump to drain base

Safe temporary side slopes of between 35° and 45° (to the horizontal) are usually achievable in gravity-drained gravels, and between 30° and 35° in drained coarse sands. Slopes which do not drain effectively under gravity will require much more site space, with side slopes as flat as 10° for stability, if porewater pressure is not reduced. In stiff clay soils, steep side slopes are theoretically feasible (short term), but sudden collapse can occur due to seepage through fine fissures and changes in pore pressure

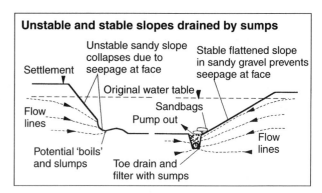

Unstable and stable slopes drained by sumps

Unstable sandy slope collapses due to seepage at face
Stable flattened slope in sandy gravel prevents seepage at face
Settlement
Original water table
Sandbags
Pump out
Flow lines
Potential 'boils' and slumps
Toe drain and filter with sumps
Flow lines

behind the slope; slope stability analysis using drained and undrained soil strengths is recommended to determine safe slope angles. If tension cracks occur, the 'critical height' of the slope is reduced.

Sump pumping can be used for dewatering trenches supported by open- or close-jointed trench sheeting depending on the soil. As the sheets have only nominal 'toe-in', the designer must ensure that inflow at the base of the trench is controlled to avoid piping, boiling or erosion leading to collapse of the trench support. If the hydraulic gradient cannot be reduced to a safe value by extending the sheeting or sandbagging at the toe, alternative dewatering, e.g. wellpointing, which causes water to flow out of and away from the excavation is indicated.

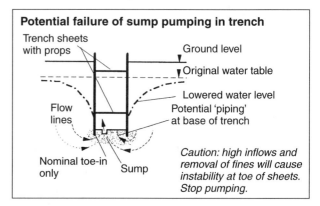

Potential failure of sump pumping in trench

Trench sheets with props
Ground level
Original water table
Lowered water level
Flow lines
Potential 'piping' at base of trench
Nominal toe-in only
Sump

Caution: high inflows and removal of fines will cause instability at toe of sheets. Stop pumping.

Installation and operation require sump excavation, usually by backhoe, placing the pump inside a perforated drum for protection and surrounding the drum by filter gravel. Adequate pump capacity and standby should be available at each sump, with due allowance for greater flow during the initial drawdown period. Where pump failure will cause rapid instability of a trench, the use of automatic water level control and switching to a standby system is desirable. In all sump pumping it is advisable to sample the water being extracted to check whether fines are being removed and provide an appropriate pea-gravel filter in the collector drains as well as around the intake. If excessive fines removal still occurs, a change in method is indicated.

If surface contaminants such as oils and cement slurry are likely to be drained to sumps then treatment of the water may be necessary before it can be discharged. Statutory, local authority and water utility regulations require the use of settlement and/or separation tanks and decanting before discharge of any groundwater to sewers or water courses – see requirement for Environmental Agency discharge consent (3.3).

Regular inspections of side slopes are advised to ensure that no water is trapped or perched which may break through in due course.

Sump pumps are usually self-priming and will handle a proportion of solids, but checks on the removal of fines which could lead to instability of the ground being drained are essential.

Electric submersible sump pumps	
Solids handling	Poor; high wear likely
Output	Flexible range of sizes from 7 m³/h for 50 mm discharge at 0.75 kW single phase to 800 m³/h for 250 mm discharge at 54 kW 3ph
Total head	Submerged pump intake; delivery from 3 m to 50 m

Air-driven centrifugal sump pump	
Solids handling	Slurries; will run dry without serious damage
Output	With air pressure at 6 bar for 50 mm discharge, up to 50 m³/h
Total head	Submerged intake; delivery up to 40 m

Hydraulic powered submersible sump pump	
Solids handling	Up to 50 mm for 150 mm pump with moderate wear over short duty; good for intermittent pumping
Output	With oil pressure at 140 bar, up to 270 m³/h with 150 mm pump
Total head	Submerged intake; delivery up to 20 m

Air-driven diaphragm pump	
Solids handling	Lightweight; useful in potentially hazardous conditions Abrasives and sludges; will run dry without damage
Output	With air pressure at 6 bar, variable up to 40 m³/h for 50 mm discharge and up to 60 m³/h for 75 mm discharge
Total head	Submerged intake; delivery up to 60 m

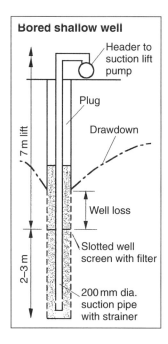

Bored shallow well

The pumps listed in the table do not rely on suction lift, as do wellpoint pumps, but operate with the pump intake fully submerged and under positive head.

The current required to start an electric pump can be up to 5 times the running current; hence the sizing of the generator is important. Phased starting is advisable for multi-well systems. The use of electric pumps on site requires installation by qualified personnel.

SHALLOW WELLS

Shallow wells are also known as 'suction wells' when the well is pumped using a suction lift pump as used for wellpointing in a 200–300 mm diameter well.

The drawdown using suction lift is limited to 6–7 m, as in wellpointing, but as a suction lift pump can be applied to each well in a multiple layout, the volume pumped from high-permeability soils is greater than from a single wellpoint line.

If suction lift and yield are inadequate, 1 m diameter shallow wells, acting as enlarged sumps to accommodate the large-diameter submersible sump pumps, can pump 800 m³/h against 10 m head, subject to well hydraulics.

Suitable for medium-term shallow dewatering in high-permeability granular soil (say, >10⁻⁴ m/s) where jetting of wellpoints is difficult. Also used in water-bearing fissured rock. Drawdown at the well depends on the method of pumping. Zone of influence and yield are calculated as for a 'deep' well.

Where multiple wells are used for dewatering large excavations and shafts, the extended discharge arrangements may disrupt congested sites.

Design of the well requires the same attention to the screen and filter as does a deep well (4.4). Slotted well screen is usually satisfactory in uniform coarse soils without the need for filter pack, provided the well is properly 'developed'. In finer soils, suitable filter and possibly geofabric 'socks' may be required around the well screen. These will reduce the inflow into the well and may make the use of the suction lift pump in each well uneconomical.

If the wells are to extract water from shallow, high-permeability aquifers, high well losses may occur due to high radial flow – i.e. drawdown outside the well may be only half that recorded in the well.

The maximum depth of a suction well is around 10 m, allowing for a sump below the pump; an internal diameter of 250 mm is required for a 150 mm suction pipe. Depth may be increased to provide pressure relief.

Spacing is 5–15 m centres for optimum drawdown with multiple suction wells.

Installation is typically in boreholes formed using light cable percussive rigs or small rotary drills producing lined diameters up to 300 mm to 10 m deep. Heavier cable tool rigs or rotary drills capable of inserting casing, as for deep wells, will be needed for the larger-diameter electric sump pumps. The heavy-duty 'hole puncher' (4.4) is not particularly effective in coarse sands and gravels.

Development of the well is usually necessary (as in the diagram) whether a filter is used or not.

Pumps for suction wells are of the vacuum-assisted, self-priming centrifugal type as used for wellpoints (4.3). The suction pipe with strainer is placed inside the screen and connected to a header pipe leading either to a central pump for several wells or to a single pump for each well – depending on well yields and critical time to failure.

Operation is similar to wellpoints, with a valve on each suction pipe tuned to maintain water level at the required drawdown in each well.

The environmental precautions noted in 3.3 must be observed during installation and operation.

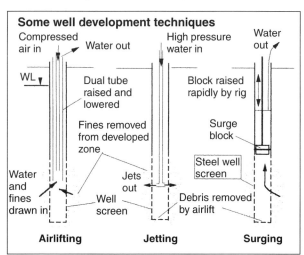

Some well development techniques

Airlifting — Jetting — Surging

4.3 Wellpoints

A wellpoint system for groundwater control is one of the most versatile methods of dewatering. It comprises a series of small-diameter pipes (50 mm) with an inlet screen (0.5–1 m long) on the lower end, each of which is inserted into the soil to a depth of 6–7 m at closely spaced centres (1–3 m). The wellpoints are connected to a header pipe at ground level and groundwater is extracted by a centrifugal pump (usually vacuum assisted) with a practical maximum lift of 7 m.

For excavations deeper than the suction lift limit, a multi-level wellpoint installation can be used on benches in the side slopes where there is sufficient space available. The economics of using deep wells or a combination of wells and wellpoints (example in 3.2) should be considered in these circumstances.

Suitable for dealing with large flows from coarse gravels and low flows from sands (k between 5×10^{-3} and 5×10^{-6} m/s for economic operation) alongside open excavations and trench works. The zone of influence is developed parallel to the line of wellpoints.

If additional vacuum is applied to a closed wellpoint system, useful drawdown can be achieved in fine soils with k down to 1×10^{-6} m/s.

The zone of influence of the drawdown will depend on soil permeability and may be estimated from 'slot' equations for inflow and yield (Cashman and Preene, 2001) – steep inflow gradient at the wellpoint in fine soils, flatter in coarse. For narrow trench excavations up to 5 m deep with trench sheeting or sheet piles, wellpoints are usually necessary on one side only – provided that the soil is permeable enough to allow a wide, flat zone of influence to develop. For wide trenches, a row of wellpoints may be needed on each side. This may apply to trenches with side support or safe side slopes.

Note: for all excavations over 1.2 m deep, side support or safe side slopes are mandatory.

Multi-level wellpoints on benches in side slopes, possibly with auxiliary sump pumping or deep wells, are generally necessary for wide excavations deeper than about 5 m. Adequate space must be available for safe side slopes.

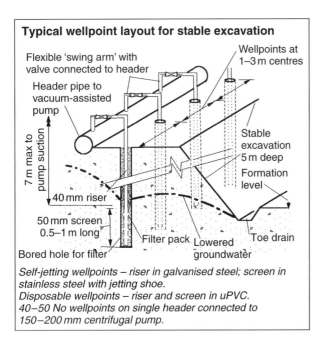

Typical wellpoint layout for stable excavation

Flexible 'swing arm' with valve connected to header

Wellpoints at 1–3 m centres

Header pipe to vacuum-assisted pump

7 m max to pump suction

40 mm riser

Stable excavation 5 m deep

Formation level

50 mm screen 0.5–1 m long

Bored hole for filter

Filter pack

Lowered groundwater

Toe drain

Self-jetting wellpoints – riser in galvanised steel; screen in stainless steel with jetting shoe.
Disposable wellpoints – riser and screen in uPVC.
40–50 No wellpoints on single header connected to 150–200 mm centrifugal pump.

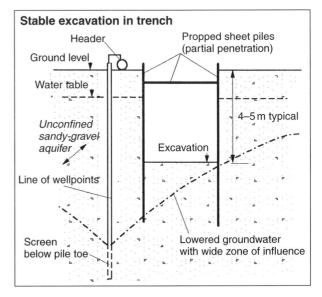

Stable excavation in trench

Header

Ground level

Water table

Propped sheet piles (partial penetration)

Unconfined sandy-gravel aquifer

4–5 m typical

Excavation

Line of wellpoints

Screen below pile toe

Lowered groundwater with wide zone of influence

Design is usually only nominal for simple trench excavations 2–3 m deep – say, 50 mm wellpoints at 1.5 m centres. For large or complex excavation, groundwater inflow into the excavation and the required drawdown should be estimated as indicated in 3.2, based on a detailed conceptual ground model.

Yield from a single 50 mm diameter wellpoint screen installed in a 200 mm diameter hole with designed filter pack will be around 1 l/s in clean sandy gravel and 0.2 l/s in fine sand, both with a vacuum-assisted centrifugal pump. The range of yield from 40 No wellpoints on a typical length of header pipe will be between 20 and 150 m³/hr. (Note 'self-jetting' wellpoints deliver only 75 per cent of these estimates.)

Spacing of wellpoints depends on soil permeability and particle size – the greater the permeability, the closer the spacing (e.g. in coarse gravel 0.3 m centres and 2.5 m centres in clean fine sand for nominal 3 m drawdown). Spacing should be based on the most permeable stratum to be dewatered.

Installation of 50 mm diameter wellpoints along the excavation is by jetting or drilling techniques.

Self-jetting wellpoints are simplest to install and are effective in silts, sands and loose medium gravel. Installation requires a large volume of water at high pressure (typically 70 m³/h at 10 bar). Jetting rate is around 50 No per day.

For long trenches (>150 m, say) the wellpoints will be installed only as needed to ensure that the drawdown is produced ahead of the excavation. The wellpoints will then be extracted on completion and 'leapfrogged' forward (the 'progressive system', as in Somerville, 1986).

Long risers in excess of the 6–7 m suction lift may be installed for deep excavations, and the header pipe progressively lowered as the groundwater is lowered.

Drilling to 7–8 m with a simple water jetting tube is the usual method of installing wellpoints which require a filter. Duplex drilling and casing (11.1) may be necessary to penetrate hard lenses. The hole puncher (see 4.4 for precautions) is a larger, more powerful version of the jetting tube used in hard cemented formations (say, SPT *N-value* 30–60) for a hole up to 300 mm diameter 15 m deep.

A continuous flight auger is useful for penetrating clay layers and for installing sand drains to drain water perched on clay layers into the pumped zone.

The use of horizontal wellpoints (4.6) could be considered where clay lenses interfere with vertical flow to the wellpoint.

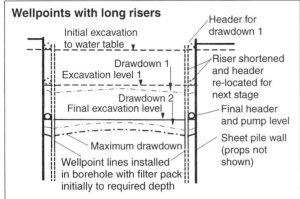

Wellpoints with long risers

Initial excavation to water table

Header for drawdown 1

Drawdown 1
Excavation level 1

Riser shortened and header re-located for next stage

Drawdown 2
Final excavation level

Final header and pump level

Maximum drawdown

Sheet pile wall (props not shown)

Wellpoint lines installed in borehole with filter pack initially to required depth

Suitable for thick layer of permeable soil.
Pumping initially from suction header main at water table level to lower groundwater for excavation to say 5 m.
Riser pipes reduced in sequence, second header connected to risers and second pump installed at excavated level to lower groundwater for next 5 m of excavation.
Pumping must be continued at initial level on reducing number of risers until next level of risers operational.

Filter pack (graded granular material) around the wellpoint is not necessary in coarse sands, but sharp sand filter should be placed in all finer soils or layered soils to prevent removal of fine particles while pumping. Development of each wellpoint to produce a natural filter in the soil around the wellpoint and filter pack is not practical (some development will occur with self-jetting wellpoints). If the discharge water contains significant quantities of fines after a few hours of pumping then action is required to improve the filter.

the specialist operator will tune the system by adjusting the flow valve at each take-off point on the header.

Once steady-state drawdown conditions have been achieved, some of the upper wellpoints in a multi-level system may be switched off for economy. Precautions for standby are needed if the excavation is subject to flooding. Regular checks should be made to ensure that:

* the tops of excavated slopes are not overloaded with plant or spoil
* uplift of the formation and removal of fines are not occurring
* seepage from side slopes is controlled
* drawdown and discharge are as required
* safety procedures and environmental requirements are enforced.

Biofouling, i.e. clogging of the wellpoint due to the action of iron-reducing bacteria (bacteria which assimilate dissolved iron and convert it to ferrous/ferric ions for energy), is not a significant problem for short-term wellpoint operation provided that the iron concentration is less than 10 mg/l (Powrie et al., 1990). It is good practice to check at least every few months to ensure that the well screen is not blocked to the extent that the excavation may be compromised, and regular back-flushing of the wellpoint is recommended. Disposable plastic wellpoints may be useful in aggressive groundwater conditions.

Pumps

An important requirement of a wellpoint pump is the ability to handle entrained air without damage to the pump and vacuum unit. Centrifugal pumps can handle air and some solids, but the filter around the wellpoint should ensure that, apart from a short development period, fines do not enter the system.

These pumps are also used as very effective sump pumps where the 100 mm pump is rated as capable of handling large solids; high-head, high-discharge centrifugal units will cope with <30 mm solids.

There are a number of factors which affect the suction performance of a pump for sumps, wellpoints and shallow well duty, reducing the theoretical lift of about 10 m to a practical lift of 6–7 m. These include:

* flow characteristics through the pump
* dry running causing overheating of the vacuum unit and 'cavitation' wear in centrifugal pumps. A flooded suction will maintain performance.

Centrifugal vacuum-assisted self-priming pump (diesel or electrically powered)	
Solids handling	Good, depending on pump size
Air handling	Up to 25 l/s
Output	160 m^3/h at 100 mm nominal rating to 700 m^3/h at 200 mm rating
Total head	Maximum suction lift 9 m; delivery from 20 to 40 m

Double-acting piston pump (diesel or electrically powered)	
	For low-flow, low-lift applications; not self-priming; poor vacuum
Solids handling	Poor, requires clean water
Output	90 m^3/h with 7.5 kW motor for 125 mm discharge
Total head	Maximum suction lift 4 m; delivery from 10 to 15 m

A header pipe of 150–200 mm diameter up to 120 m long collects the water from each wellpoint riser and is then connected to the vacuum-assisted, self-priming centrifugal pump.

Standby pumps should be connected to the header so that immediate transfer can be made in the event of pump failure. If the pumps are electrically powered from the mains grid, it is advisable to provide standby generators (this may be an insurance requirement).

Additional vacuum may be provided to improve the wellpoint yield and drawdown in fine soils by reducing the atmospheric pressure in the soil, either by effectively sealing the top of the filter pack with a clay plug or by connecting additional vacuum capacity to the header pipe. As gravity drainage in such fine soils will be slow, some water may remain in pores, and while excavated side slopes in these conditions can be stable, a true zone of influence may not be achieved.

Operation

Drawdown at the wellpoint should not be below the top of the screen to avoid pulling in air and causing difficulty in maintaining vacuum at the pump. To prevent air entry,

Wellpoint line to trench excavation

4.4 Deep Wells

All wells for dewatering could be considered relatively shallow when compared with water supply wells. Deep wells for construction dewatering are pumped with air-lift (inefficient for the long term), a vertical shaft-driven pump or high-head submersible pump installed in a borehole and are able to produce greater drawdown than possible with shallow suction wells or wellpoints.

Suitable for long-term dewatering to produce deep drawdown for large excavations in a wide variety of ground conditions ($k > 10^{-6}$ m/s). Generally best suited to relatively deep, thick aquifers of gravels to silty sands, they will cope with thin intervening beds of low-permeability strata in the main aquifer – particularly if the well filter pack intercepts these beds and the well is properly developed. They also provide under-drainage of overlying, less permeable soil into the pumped permeable stratum and pressure relief below the confining clay layer.

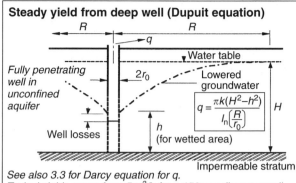

Steady yield from deep well (Dupuit equation)

$$q = \frac{\pi k (H^2 - h^2)}{\ln\left(\frac{R}{r_0}\right)}$$

See also 3.3 for Darcy equation for q.
Typical yields range from 5 m³/h for a 150 mm diameter well to 500 m³/h for a 600 mm well.

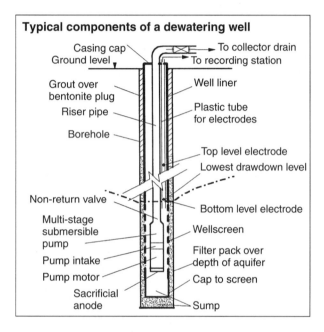

Typical components of a dewatering well

Casing cap
Ground level
To collector drain
To recording station
Grout over bentonite plug
Well liner
Riser pipe
Plastic tube for electrodes
Borehole
Top level electrode
Lowest drawdown level
Non-return valve
Bottom level electrode
Multi-stage submersible pump
Wellscreen
Pump intake
Filter pack over depth of aquifer
Pump motor
Cap to screen
Sacrificial anode
Sump

Design of wells is briefly outlined in 3.2 and 3.3. It is essential that a detailed conceptual ground model is developed from a comprehensive ground investigation. Pumping test data – preferably with distance/drawdown/time results – are important. Some basic steps to follow:

- determine drawdown needed at the well to suit the excavation geometry
- calculate inflow into the excavation (Preene *et al.*, 2000; Cashman and Preene, 2001); use flow nets to assess inflow in large areas and to check potential for high inflow velocity causing boiling or piping
- calculate the yield of a trial well (say 150–300 mm diameter screen plus filter) and wetted length
- calculate the number of wells and spacing, and check interaction (or superposition) of zones of influence for wells at <20 m centres
- design the filter pack and screen slots to ensure that fines from the aquifer are not drawn into the well
- check the entry velocity at the well (inflow rate/slot area <0.03 m/s to avoid differential levels inside and outside the well)
- check the pump sizes available to suit the screen/liner
- ensure sufficient working space for excavation and side slopes
- repeat calculations to obtain optimum design and check cost-effectiveness.

Well depth must be adequate to provide sufficient wetted area of intake in the aquifer to produce the interactive drawdown and yield. For most civil engineering requirements, the maximum well depth is around 60 m, with 20 m being typical. The well may penetrate for a limited distance (1–2 m) into an impermeable layer below the pumped aquifer to act as a sump for fines.

The pump should be set at a level in the well where water will circulate to keep it cool.

Spacing of wells is frequently dictated by the site layout but 10–30 m centres is usual; deepening wells may improve drawdown in areas where wells cannot be sited. Check that increasing drawdown does not lead to settlement under adjacent buildings – requiring either a cut-off barrier or a recharge system.

Well screens for the required intake wetted length are designed to allow gravity flow of water into the well, without disturbing the unconsolidated strata being pumped. A blank well liner supports the remainder of the hole. Materials may be plastic or steel (occasionally stainless) depending on the duty required. Mixing different metallic materials will set up corrosion cells.

The volume of water transmitted by the screen depends on the open area of the screen – the size and number of slots. The open-wetted area provided should be greater than required for steady flow by a factor of 1.2–1.5. Typically the slot should be no larger than the D_{10} particle size of the filter pack, with the open area being 5 per cent of the screen cylinder area for fine soils and 25 per cent for coarse soils in order to avoid well losses. For layered soils, the D_{10} of the finest soil from the combined grading curves may be used as the basis for design, but check excessive head loss through the filter.

The openings in the screen are produced by cutting, drilling or burning slots. The precision-made wire-wrapped Johnson 'continuous slot' screen developed initially for oil wells (Driscoll, 1986) allows controlled high inflow and may avoid the need for filter pack in coarse to medium soils.

If the wetted area is reduced during drawdown so that the required flow cannot be achieved from the aquifer, additional wells or larger diameters are needed.

Filter pack (graded granular material) is usually required between the screen and the soil to ensure that only fines-free water enters the well. Alternatively, layers of geotextiles can be wrapped around large-slotted well screen to control the entry of fines. Design rules using the particle size distribution curve of the aquifer to be protected are given in BS 8004: 1986. Note that fines may have been lost during the sampling of the *in situ* soil; also that these rules have to be relaxed for silty aquifers.

While it is difficult to layer different filters in the borehole to match different soil gradings at different depths in the aquifer, it is feasible to have different precision screen sizes in the same well.

Installation of wells is usually at the top of excavation slopes or outside sheeted trenches. Discharge pipes and collector drains should be carefully sited to avoid damage and congestion.

Well drilling equipment includes cable tool percussion rigs, rotary drills using direct or reverse circulation (11.1), duplex casing and hole puncher jetting rigs. Depending on the equipment used, completion of one well per day to, say, 10 m is reasonable. In all cases the driller should take samples and log the strata to give confirmation of the soil to be screened.

> **Caution**: the hole puncher is very heavy; it operates on a crane which uses a free-fall drop-hammer to drive a steel tube. An internal jetting pipe, using compressed air in conjunction with high-pressure water, throws out material from the ground. Personnel must stand clear and adjacent equipment must be protected.

The essential requirements are to produce a stable hole for the insertion of the screen and filter over the relevant aquifer and to avoid disturbing the unconsolidated soils (or, when rock drilling, to avoid blocking water-bearing fissures).

Screen and permanent lining are connected in lengths as required and lowered into the hole with centralisers. The filter gravel is then carefully placed in the annulus around the screen using 50 mm tremie pipes to avoid segregation of particles, and any temporary casing supporting the aquifer is withdrawn. The top of the filter should be sealed to prevent surface leakage into the filter and assist in producing a degree of vacuum in the capped well liner.

Development (or cleaning) of the well is carried out as soon as possible after installation to remove drill debris and optimise the grading in the aquifer immediately around the screen to deliver fines-free water. Typical development methods (see 4.2) include surging of water into the well and jetting with a horizontally directed water jet; pumping and air-lifting are less effective. Specialised techniques such as 'acidising' in chalk and, on rare occasions, hydrofracture in rock, may be needed to optimise the yield.

If clean water is not produced within about 12 h, then the filter and screen may not be effective and a second string may have to be placed inside the original – reducing the available diameter for the designed pump.

The disposal of the muddy and sandy water must be controlled through settling tanks or treatment plant before being discharged to the approved route (Masters-Williams *et al.*, 2001).

Additional vacuum may be applied to improve yield in fine soils by sealing the well with bentonite grout and then applying a vacuum pump directly to the well liner. Note that with time, the initially improved yield may be lost as more fines are sucked into the filter pack.

Operation of most deep well pumps requires an electric power supply from the mains grid or three-phase generators. On long-term projects it is advisable to operate on the mains with automatic cut-in of generators in the event of power failure – this may be an insurance requirement. Diesel-operated shaft-drive pumps are used where electric power is unavailable, but require more attention to correct installation.

The provision of adequate standby pumps ready to be installed immediately a failure occurs is desirable but not always cost-effective. The time from failure of wells to the collapse of the excavation due to ingress of water should be calculated. Adjacent wells should have capacity to take over the duty of a failed well for a short period to allow remedial action to be taken.

Level-monitoring electrodes are placed at maximum and minimum levels in the borehole so that the pump does not run dry and the designed drawdown is maintained. The electrode spacing and discharge rate should be such that no more than six starts of the pump occur per hour. In addition careful manual tuning of the surface discharge valve will help to maintain steady pumping. When the designed drawdown has been achieved, it may be possible to switch off selected wells – subject to checking ground failure times.

Environmental precautions as in 3.3 are essential.

Biofouling or encrustation on the screen due to long-term operation (as noted for wellpoints) will not normally be a problem for high yield wells with wide screens in coarse soils. Precision screens in fine soils will require back-flushing regularly to remove fouling if iron concentrations are >5 mg/l. Steel piling adjacent to long-term dewatering wells will contribute to blockages caused by the iron-related bacteria.

Pump installations range from a simple air-lift pump for short-term (expensive) use, to top-drive vertical-shaft pumps and submersible electric pumps, including sump pumps in large-diameter wells. It is not good practice to use pumps of higher output than needed. Large line shaft-driven pumps require accurate vertical placing in vertical liner and screen if the shaft is not to vibrate.

The most versatile pump is an electrical submersible borehole pump installed in a well on a steel riser pipe fitted with non-return valve, discharge valve and pressure gauge.

If high output at high-delivery head is required, a large-diameter borehole 600–1,000 mm will be needed to accommodate a large electrical sump pump similar to that for sump duty (4.2).

Regular cleaning of the pump intake is essential.

Helical screw rotary displacement electric shaft-drive pump

Simple to install to depths of 15 m; maintenance convenient; can handle some fines
Note: the large pumps are up to 3 m long and require clutches on the top-drive motor
Output range: For 100 mm well screen, 1 m³/h at 0.5 kW to 3 m³/h at 1 kW for total
            head of 15 m
            For 150 mm well screen, 50 m³/h for total head of 15 m at 7.5 kW 3ph
            to 30 m³/h for total head of 120 m at 22.5 kW 3ph

Multi-stage electric submersible pump

Relatively simple to install; reliable if operated and maintained as recommended;
must be completely submerged when operating; will not handle fines
Output range: For 100 mm well screen, 0.5–10 m³/h for total head of 10–200 m at
            0.5–2 kW 3ph
            For 250 mm well screen, 90–140 m³/h for total head of 10–75 m at
            7.5–50 kW 3ph

Electric submersible pumps

4.5 Eductors

The eductor or 'ejector' is essentially a jet pump based on the venturi principle, placed on a riser at the bottom of a small diameter well. It is not dependent on suction lift. The jet operates at a supply pressure of >7 bar to produce a vacuum of around 1 bar in the venturi, which draws in the groundwater to be returned to the low-pressure surface collector pipe.

Suitable for removing small quantities of water from low- to medium-permeability ($k = 10^{-7}$ to 10^{-5} m/s) fine-grained soils. Eductors will operate satisfactorily to dewater soil and control porewater pressure at a practical maximum depth of 50 m and are useful as an alternative to multi-stage wellpoints in medium-permeability formations.
Not recommended where biofouling will clog the nozzle or where fines cause excessive wear.
Not energy-efficient – costly to run.

Design of inflow to the system is calculated as in 3.2 and 3.3, based on the detailed conceptual ground model and relevant soil parameters.

Types of eductor include:

- a single-pipe eductor installed in a well or wellpoint above the intake screen with the well liner acting as the return pipe from the eductor nozzle
- a two-pipe system with separate high-pressure injection pipe to, and low-pressure return from, the eductor placed above the intake screen in a 100–200 mm diameter well.

The eductor manufacturer should provide specific performance curves for each type of eductor giving the water-supply pressure and volume needed to produce the required induced flow through the nozzle system (otherwise field tests on the particular eductor are needed (Miller, 1988)).

Layout is usually on ring mains around the top of the excavation; in lines for porewater reduction under embankments.

Drawdown up to 30 m is achievable at the usual injection pressures and to 50 m with higher pressure – although the improvement in well discharge falls off rapidly as cavitation occurs at the higher pressure (Powrie and Preene, 1994). The eductor system can therefore effectively replace multi-stage wellpoints for deep excavations in low-yield fine soils.

Yield of a single eductor is between 0.1 and 0.4 l/s from low-permeability soil depending on the manufacturer's performance data. Some suppliers provide various diameter venturi/nozzles to alter yield.

Well screen and filter are designed to suit the strata being drained, as for deep wells. It is essential to prevent entry of fines causing wear in the venturi/nozzle.
Water level in the well will usually be lowered to just above the nozzle level at optimum operation. If the nozzle is within the screen the wetted area of inflow will be reduced. This will be important when drawing down groundwater to rockhead for excavation.

Development of the well to remove fines should be carried out before installing the eductor.

Spacing is 1–5 m depending on permeability, yield and drawdown needed. The superposition assessment is critical to cost-effectiveness. In cases where perched water has to be drained with eductors, spacing, filter diameter and depth of eductors must be carefully selected. Attempting to improve vertical drainage by punching through clay layers with sand drains is unlikely to be effective in the short term.

The number of eductors on the high-pressure line depends on total length of injection pipe, friction losses (which can be high) and pumps available. In the most favourable conditions up to 60 eductors at 1 m centres can be designed on each header.

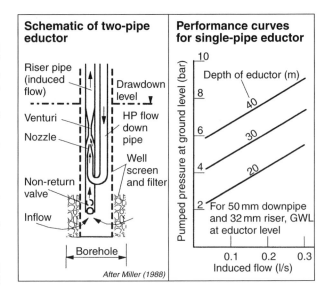

Schematic of two-pipe eductor	Performance curves for single-pipe eductor

After Miller (1988)

Additional partial vacuum (say, 0.5 bar) can be developed by the eductor drawing in air from the surrounding filter pack by sealing the top of the filter with a bentonite seal. This will improve soil stability as a result of reducing porewater pressure in fine soils (Powrie et al., 1993).

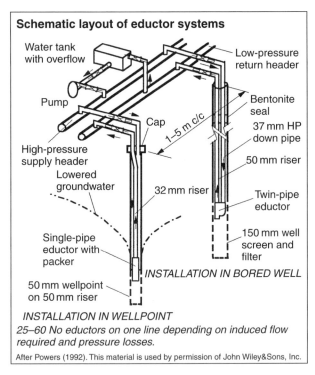

Schematic layout of eductor systems

INSTALLATION IN BORED WELL

INSTALLATION IN WELLPOINT
25–60 No eductors on one line depending on induced flow required and pressure losses.

After Powers (1992). This material is used by permission of John Wiley&Sons, Inc.

Installation for most applications requires the drilling of a cased hole to the required depth capable of accommodating a 100–200 mm well screen and filter pack for the two-pipe eductor or a 50 mm single-pipe wellpoint-type eductor and filter.

Drilling methods are as for wellpoints and wells, depending on the depth and diameter required. Environmental precautions during drilling and discharge as noted in 3.3 should be observed.

Pipework duty must be matched to the high supply pressures required. The high-pressure supply pipe is usually 100 mm in diameter and the low pressure return 150 mm. On large sites with non-rectangular excavations the pipework and pumping system can be complex, requiring expert installation and supervision.

Operation of the eductor will continue even while pulling in air when drawdown is near nozzle level. Trimming of the high-pressure water supply is usually not necessary once the system has been tuned initially. Pressure gauges should be provided on the supply line.

Supply pumps must match the requirements of the nozzle to deliver the eductor-induced flow. The number of eductors per pump depends on depth and yield needed; several pumps can be connected in parallel to the supply line. For large systems several smaller pumps give better flexibility than a single large pump.

Water has to be supplied at pressures of 8–15 bar at flow rates of 1–2 m³/h to each eductor, usually by positive displacement high-pressure pumps or high-speed, multi-stage centrifugal pumps.

Increasing supply pressure above the nozzle design requirements will generally not improve induced flow.

The pipework layout can be complex and significant friction losses can occur.

Monitoring of drawdown or reduction in pore pressure using eductors requires accurate piezometer measurements – simple standpipes will not respond adequately in fine soils. Pressure changes in the supply line will indicate problems at the nozzle – possibly requiring a time-consuming replacement.

Groundwater lowering system with ring of interacting eductor wells

Supply and return mains
Radius of influence R
one line pumping
Equivalent well radius r
Original water table
Formation
Lowered groundwater for ring of eductors interacting
Well or wellpoint
Cone of depression for one line operating
Reduced entry gradient (increased with additional vacuum)
Unconfined silty sand aquifer
Eductor body

At centres > 20 m adjacent deep wells will have marginal effect on each other. Yield of eductors at 1–5 m centres will be significantly influenced. Design of multi-well system (Somerville, 1986) requires the application of the methods in 3.2 and 3.3 to detemine:

- 'equivalent radius' of excavation, zone of influence, drawdown required at centre and at well (parabolic relationship), hence inflow into excavation;
- wetted length of inlet to well and intake level of the eductor or pump to produce drawdown;
- yield and number of pumps/eductors required.

Cavitation is not particularly detrimental to eductor performance at normal injection pressure, but wear by drawing in fines is. Regular checks are therefore required on the outflow from the eductor (using a measuring weir in the overflow of the collector tank) to determine whether wear at the nozzle/venturi or biofouling is reducing yield. Pipework must also be inspected for signs of fracture as burst pipes under the high operating pressures can cause surface erosion and flooding.

Standby pumps on each eductor line are essential if critical time to slope failure is short. The power required per litre of water delivered is higher than for wellpoints and deep wells and therefore not efficient for large inflows.

Biofouling will seriously affect low-flow eductors with low iron concentrations (<5 mg/l), requiring cleaning of the eductor every few months. With concentrations above 10 mg/l an eductor system may need cleaning so often that using eductors is not appropriate. Encrustation will also cause clogging in hard (calcium carbonate) groundwater.

High-speed single-stage centrifugal pump (diesel)	
Output	500 m³/h for 150 mm discharge at 8 bar total head
Six-stage centrifugal pump (electric powered)	
Output	110 m³/h max. at 11 bar total head; 45 kW 3ph

Case: Dewatering for earth dam in Brunei
In order to accelerate the consolidation of the soft clay foundation to the shoulders of a 20 m high earth dam, vacuum eductors were installed at rockhead (R G Cole et al., 1994). Initially, 128 No 50 mm diameter single-pipe eductors at 1 m centres in 200 mm boreholes were installed in lines near the upstream and downstream toes of the dam, with provision made for two additional lines to 45 m deep within the wide shoulders if geological conditions warranted. The filter pack over the length of the riser was plugged at the top with bentonite to develop a modest vacuum.

Piezometers were installed adjacent to the riser in selected holes.

Pumping stations were provided at the mid-point of each line comprising two covered weir collector tanks, one operating pump and one standby on one line and two operating and one standby pumps on the longest line, all rated at 145 m³/h at 7 bar for ease of interchange and maintenance. Typical yield per eductor was 0.45 l/s initially, reducing to 0.03 l/s for the operating period. Groundwater was reduced by more than 20 m, close to rockhead.

Installation took 4 months and the system operated continuously for 27 months.

Line of eductors

Eductor pump set-up

4.6 Horizontal Wellpoints

A flexible perforated pipe is placed at the bottom of a deep slit trench and pumped with a wellpoint pump to lower groundwater. The method is no longer popular in UK as the large purpose-built trenching machines to lay deep drainage pipes to make full use of the 6–7 m suction lift of wellpoint pumps are not routinely available. Agricultural drainage trenchers with depth capability of 3–4 m have been used for dewatering for pipelines; backhoe excavators are not efficient for pipe installation.

Suitable for long pipeline runs to dewater silty sands and medium gravels.

Not recommended in clayey soil as the drain will become clogged with slurry; but useful to cut through thin clay aquiclude.

The effective drawdown depth is limited by suction lift, as for wellpoints. Drains are best positioned directly under trench to be dewatered and are useful as an alternative to sand drains in accelerating consolidation of large areas of soft soils, particularly if additional vacuum pumping is applied.

Benefits include clear working area and rapid installation.

Design is nominal, based on the capacity of the drainage pipe and pump. A 150 mm rated wellpoint pump will drain 100 m length of 150 mm diameter drain at a depth of 4 m. Filter materials are as for wellpoints.

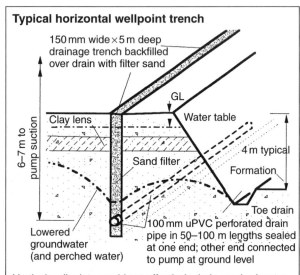

Typical horizontal wellpoint trench

150 mm wide × 5 m deep drainage trench backfilled over drain with filter sand

GL

Clay lens

Water table

6–7 m to pump suction

Sand filter

4 m typical

Formation

Lowered groundwater (and perched water)

Toe drain

100 mm uPVC perforated drain pipe in 50–100 m lengths sealed at one end; other end connected to pump at ground level

Vertical wellpoints would not effectively drain perched water above clay lens.
For a pipeline trench the horizontal wellpoint is placed on the line of the trench.
Typically: 1 No centrifugal pump for 200 m of 80 mm drain
1 No centrifugal pump for 100 m of 100 m drain;

Installation involves a drainage trencher laying the perforated plastic drain pipe to a maximum depth of 7 m and simultaneously backfilling over the pipe with the gravel filter. Pumps are connected to the drain, which is brought up to the surface every 100–200 m.

Outputs vary according to depth, ground conditions and type of trencher. An agricultural unit will place up to 1,000 m of drain 1 m deep per day; a specialist trencher capable of digging a 6–7 m deep trench in long runs may achieve over 500 m per day.

Operation of a horizontal wellpoint system by self-priming, vacuum-assisted centrifugal pumps requires little attention, except for tuning to ensure that the vacuum unit is not overworked due to pulling in air.

Back-flushing of the drain may be needed if output falls due to silting or biofouling.

VACUUM DRAINAGE

A vacuum can be applied to the sand-filled wellpoint trenches, with a sealed drainage blanket at the surface, to improve dewatering and consolidation of soft compressible soils.

A vacuum wellpoint pump is placed on each 100–200 m long trench, and with a plastic sheet to form the seal, a vacuum of 0.5 bar is feasible. With separate vacuum pumps the vacuum can be increased to 0.8 bar over a considerable area, and the wellpoint pump operated intermittently. A separator should be placed in the vacuum line to collect air and water drawn in.

The area of drain in contact with the soil is much larger in this layout than that available in vertical drains (6.1) and consolidation can be much quicker.

The addition of fill to compensate for the settlement can be added as pre-load to accelerate consolidation (6.2).

Monitoring of settlement, piezometric levels and vacuum achieved will indicate effectiveness, as for regular wellpoints.

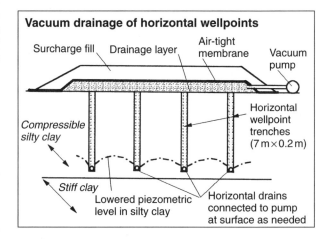

Vacuum drainage of horizontal wellpoints

Surcharge fill Drainage layer Air-tight membrane Vacuum pump

Compressible silty clay

Horizontal wellpoint trenches (7 m × 0.2 m)

Stiff clay

Lowered piezometric level in silty clay

Horizontal drains connected to pump at surface as needed

HORIZONTAL WELLS

With the development of small-diameter directional drilling techniques and computer-controlled steering systems, horizontal wells or drains can be installed at greater depths than by using a trencher, for lengths of 1,500 m up to 1,000 mm diameter. In addition to targeting specific geological strata, they can be used to clean up contaminated land by:

- intercepting and drawing off contaminated leachate
- air sparging and extraction of vapour and water under vacuum
- injecting nutrients for *in situ* bioremediation
- intercepting leaks from the base of landfill and contaminated groundwater plumes.

The well may be constructed with a 'blind end' in the ground – difficult to install the liner and screen. By continuing drilling in an arc to the surface the lining and screen can be inserted by back-reaming and pulling.

If water is to be extracted by suction lift, the depth of the horizontal drain will be limited. Flexible air-lift systems will deal with high volumes of water but are expensive.

The wells for remediation of contaminated land should be designed by qualified geoenvironmental engineers and installed by experienced personnel with the appropriate equipment.

Caution: contaminated water recovered from landfill will require special measures and treatment prior to disposal (Hardisty *et al.*, 1996).

4.7 Relief Wells

Suitable for controlling excess porewater pressure or artesian pressure in a confined aquifer below an excavation using designed wells but without using water extraction pumps; also known as 'passive' wells. Relief wells can deal with aquifer permeabilities from 5×10^{-5} to 1×10^{-2} m/s and relatively high inflows.

They are used for temporary groundwater control and permanent under-drainage and seepage control for deep cuttings, dams and weirs.

They can successfully control tidal conditions in a cofferdam when combined with sump pumping and excavation management.

Not recommended where there are weak lenses at the base of the excavation which may heave before the pressure is relieved or where the water discharging on to the formation causes serious softening.

Vertical sand drains to accelerate consolidation of foundation soils (6.1) are a form of pressure relief.

Ground investigation where artesian conditions may exist should allow for provision of accurate piezometer readings over wet and dry seasons in order to anticipate the need to relieve such pressures, both during construction and long term (McKenna *et al.*, 1985).

Design equations to estimate flow into a relief well system are usually based on artesian conditions in a confined aquifer with either fully or partially penetrating 'equivalent' wells (Cashman and Preene, 2001). For cofferdams and large excavations flow nets are desirable to assess stability. In all cases the discharge elevation must be taken into account.

Spacing will be 3–10 m in temporary conditions, but where large-diameter wells (say, 600 mm) are used under dam drainage blankets spacing can be greater. While one or two wells may adequately relieve pressure below a shaft during construction, multiple wells will generally be needed, with analysis of the interaction (superposition) principle noted for deep wells and eductors.

Screen and filter are required for relief wells in high-permeability aquifers, as for deep wells, for optimum efficiency. For permanent pressure-relief systems, the use of stainless steel screen may be needed, but plastic screens and liners are generally acceptable. For lower-permeability applications the borehole may simply be filled with the appropriate grading of rounded gravel.

Yield using the Darcy equation gives an estimate of the capacity of a sand- or gravel-filled well, with the hydraulic gradient taken as 1 (Powers, 1992). Typical discharges range from 3 l/h for a sand-filled 100 mm diameter well to 3,600 l/h for a pea gravel-filled 300 mm well, depending on the degree of smearing of the borehole sides.

Prevention of heave in excavation by relief wells

Piezometric level in lower sand
Original water table
Drainage blanket
Pump out
Stable slope (gravity flow to drain and sump)
Flow
Sand
Relief wells with screen and filter
Silty clay
Lower sand
Uplift pressure relieved

Sumps and drainage blankets are required for temporary works, and collecting chambers, mitre drains or headworks are required on large hydraulic structures to control the up-flow from the well system.

Installation will vary depending on whether wells are for permanent or temporary works.

Drilling methods are as for 100–500 mm diameter deep wells. If the piezometric pressure is above ground level, upstanding casing is needed, requiring a drill platform to handle tools in and out of the hole.

The wells may have to be drilled prior to excavation to avoid pressure on the base at all times. If the well is lined and screened, it will be necessary to break out the liner as excavation proceeds and provide pumping to remove the discharge below the artesian level.

If the drainage blanket is used as the drilling platform it will be necessary to clear away any potentially clogging fines from the drill arisings, particularly in permanent drainage conditions under a dam (see also 6.1).

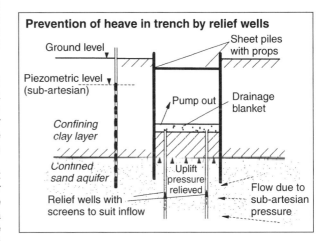

Prevention of heave in trench by relief wells

Ground level
Sheet piles with props
Piezometric level (sub-artesian)
Pump out
Drainage blanket
Confining clay layer
Confined sand aquifer
Uplift pressure relieved
Relief wells with screens to suit inflow
Flow due to sub-artesian pressure

Development of a well including a screen and liner is desirable to allow maximum controlled flow into the well. For wells without a screen, development is not feasible as the filter material is tremied into the hole as the temporary casing is withdrawn.

Pumping tests of a screened well are useful to confirm conditions and application of the observation method for large excavations or permanent pressure relief.

Operation requires adequate sump pumping capability or gravity drainage of discharge water from the filter blanket so that placement of foundation structure in the dry is not impeded.

Temporary pressure relief must continue until there is adequate weight from the permanent structure to resist heave and uplift.

Ideally, on completion of temporary relief wells, they should be effectively backfilled and sealed. For sand-filled wells this will require removal of an adequate depth of the sand and injection of a plug of grout or concrete to counter uplift, with a factor of safety on the plug weight. For screened wells grout is injected as the screen is removed. Care is required not to contaminate the surrounding groundwater with cement slurry. Environmental restrictions will apply to the discharge from these wells.

Biofouling is difficult to deal with in relief wells backfilled with filter as back-flushing is not possible – reduction in yield is inevitable. While this may not affect short-term construction operations, if clogging due to groundwater chemistry or iron bacteria is indicated then screens should be provided for long-term use.

4.8 Recharge Wells

Recharge wells are used in construction to limit the adverse effects of the large drawdown necessary for deep excavations in urban areas by returning water to the aquifer away from the site dewatering system, possibly in conjunction with a cut-off structure.

Suitable for controlling settlement of compressible soils due to porewater changes resulting in damage to buildings and services; also for the maintenance of water supply aquifers. They can be useful in preventing movement of saline and contaminated groundwater towards the dewatered zone when used in conjunction with a cut-off system, and are also useful for disposing of discharge water in the absence of other routes.

Not recommended where serious clogging by biofouling or encrustation of the well can occur or where fines may be present in the discharge.

Cannot be used where discharge/recharge consent is not available.

Ground investigation should include pumping tests in the recharge area and chemical analysis of the groundwater to be abstracted to check for potential biofouling.

The conceptual ground model should provide detailed hydrogeological data for the areas around the dewatered site where recharge of the aquifer is to take place.

Design should ensure that the recharge area will be beyond the zone of influence of the dewatering or behind an effective cut-off. The well can be designed as in 3.2 and 3.3 to discharge to a specific high-permeability stratum, preferably where there is a deep water table. The use of numerical modelling to analyse adjacent extraction and recharge is now recommended.

Screens and filters are designed as for deep wells and should be as coarse as possible in relation to the formation into which recharge water is to be injected. The exit velocity from screen to receiving stratum should be around half the entrance velocity for abstraction wells (4.4).

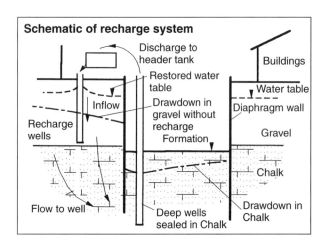

Schematic of recharge system

The number of wells required is a function of:

- potential for clogging
- the need to take wells out of service for maintenance
- the diameter and slot area for exit velocity.

Typically, 2–3 recharge wells are required for each pumping well in the same aquifer. Locations should be chosen to ensure recharge water does not re-enter the dewatering system.

Well diameter and wetted area are selected to provide the required outflow. Diameter may be smaller than an abstraction well as there is no pump to accommodate – but allowance must be made for access for periodic cleaning of the well.

Wellpoints may also be used as recharge systems, particularly in near-surface aquifers which have low acceptance rates (low transmissivity). They are designed at similar spacing to abstraction systems and adequate area for installation is needed.

Recharge trenches are rarely convenient for construction works unless shallow trenches can be located to feed shallow aquifers.

Installation is based on the standard well drilling methods and screen assemblies. The upper section of the liner should be sealed to the formation to prevent injected water rising up the filter to the surface.

Development of the well is necessary but care must be taken to avoid removal of groundwater which will have adverse effects on the structures which the recharge system is designed to protect – surging and jetting are preferred to air-lifting or pumping in such cases.

The abstracted water should be collected in a de-aeration/settlement tank prior to re-injection – usually by gravity head a few meters above water table. A down

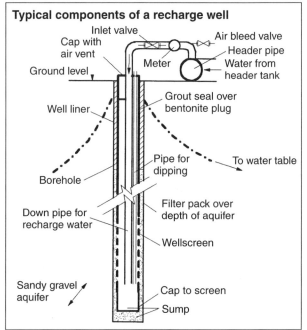

Typical components of a recharge well

pipe in each well is connected to the header pipe and tank with air bleed-off valves at high points. In complex systems care is required in calculating pipeline friction losses at low heads.

Operation of a recharge system usually takes the discharge from the dewatering system. Clean, high-quality water is necessary – without suspended fines and of acceptable chemical composition – to avoid serious clogging and contamination. This may require treatment of groundwater or the approved use of mains supply water in order to obtain the discharge consent. Where chemical dosing of the recharge water to combat encrustation and bacterial growth is indicated, environmental constraints may intervene to prevent such remedies.

Water supplied to each well should be metered and the recharged water table or confined aquifer monitored by standpipes or piezometers.

Cleaning of single wells in rotation by air-lift or pumping development techniques is acceptable once the recharge system is in full operation.

4.9 Electro-osmosis

Groundwater flow through some low-permeability silts and clays can be improved by passing a direct electric current (DC) through the soil between electrodes placed in the ground. The cations in the porewater are attracted to the cathode, thereby causing flow. The activity of the clay fraction will affect the removal of the diffused water from the soil: more water will be removed from a silty, low-activity soil than from a high-activity clay for the same applied current. However, the electro-osmotic permeability of high-activity clay may be several orders of magnitude greater than its hydraulic permeability, allowing the strength of soft compressible clay to be improved.

Electro-osmotic flow is costly to develop, hence only rarely used, and then usually in conjunction with electro-chemical stabilisation – see below.

Suitable for porewater pressure control in low-permeability, near-saturated silts and clayey soils which do not drain adequately for, say, eductor pumping methods. Shear strength adjacent to the electrodes is improved as porewater is removed at the cathode – assuming that total stress remains constant. Some strength improvement is also seen in clay due to the base exchange reaction with iron (Fe) ions from the anode forming stable ferric hydroxides in the soil.

Not recommended for fissured or varved clay – sand drains are much cheaper.

Design is based on Casagrande's work (1952), which requires assessment of the ground conductivity and application of an empirical form of the Darcy equation to determine the flow.

Direct current is applied to sacrificial steel anode rods in the soil and the circuit is completed by connection to the steel dewatering wellpoint forming the cathode, installed 1–5 m away. The water flows from the anode to the cathode at a rate dependent on the electro-osmotic permeability, soil electrical conductivity and electrical potential applied.

Yield depends on depth (typically 5 m, but 30 m has been achieved) and method of extraction – up to 0.1 l/s for eductors and 0.3 l/s for wellpoints.

Spacing of electrodes is dependent on the potential gradient applied – typically 50 V/m but possibly as high as 200 V/m initially; the higher gradients being required for sandy soils. Power required ranges from 0.5 to 1.5 kWh/m^3 of ground dewatered at currents between 25 and 100 amp; greater for small excavations.

Electrodes are usually the steel wellpoints/eductors (cathode) extracting the water and iron rods (anodes). However, various other materials and intermittent current configurations have been tested to improve electro-osmotic permeability with limited success (Mohamedelhassan and Shang, 2001).

Installation of the cathode wellpoint or eductor is as described above (4.3, 4.5) and the anode is drilled or driven into the soil depending on conditions and depth needed. The DC connections must be made by electricians in accordance with codes.

Operation for removal of groundwater is as for well-point/eductor procedures. As water is extracted the process becomes less efficient but the cost of electrical power decreases. In certain flow conditions porewater pressure may increase around the cathode and cause some softening of the soil.

Standby generators and pumps are not usually required for safe operation of the slow stabilisation process.

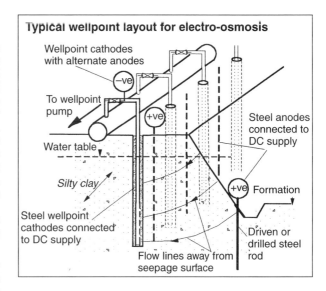

Typical wellpoint layout for electro-osmosis

Wellpoint cathodes with alternate anodes

−ve

To wellpoint pump

+ve

Water table

Silty clay

Steel wellpoint cathodes connected to DC supply

Steel anodes connected to DC supply

+ve Formation

Driven or drilled steel rod

Flow lines away from seepage surface

Problems occur with anode corrosion requiring periodic renewal, and with electrical contacts operating under high current density.

ELECTRO-OSMOSIS WITH VERTICAL DRAINS

Consolidation of soft compressible clay can be accelerated by combining prefabricated vertical drains (PVDs or band drains; see 6.1) with steel anodes and cathodes – the steel inserts to the PVDs replacing the steel wellpoints in low-permeability clay. The drains are installed with a mandrel and a drainage blanket is required at the surface.

Experimental data (Bergado et al., 2000) show that 90 per cent consolidation of soft clay using a voltage gradient of 60 V/m and reversing the polarity of the electrodes regularly (to produce more uniform drainage) was achieved 3 times quicker with PVDs and electro-osmosis than with PVDs alone.

Similarly there was an increase in shear strength at the electrodes compared with the PVDs on their own.

Higher voltage gradients produced higher strengths, but the system is likely to be very costly for most construction projects.

Electrically conductive geosynthetics are being developed as electrodes placed in reinforced earth to improve shear strength of the fill by electro-osmosis (Lamont-Black, 2001). Use of these materials as PVDs would appear feasible for electro-osmosis.

ELECTRO-CHEMICAL STABILISATION

The process described above can be used to increase the shear strength of soft clay by releasing an electrolyte (say, calcium chloride) into the soil through the anode (Farmer, 1975). This improves the release of the stabilising ions, but flow reduces with the increase in electrolyte concentration. Slow and expensive.

CONTAMINATED LAND

There is potential for using electro-osmosis to decontaminate low-permeability soils. Dissolved organic compounds could be induced to move towards the cathode, where the contaminated water could be removed for treatment.

Extracting inorganic and heavy metal contaminants is likely to prove difficult using this process.

5 GROUNDWATER CONTROL – EXCLUSION METHODS

5.1 Grout Curtains

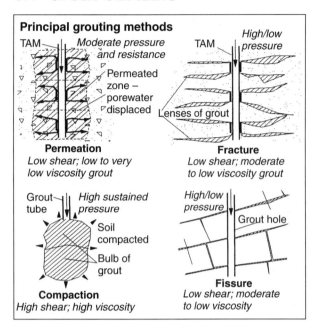

Principal grouting methods

TAM *Moderate pressure and resistance*

Permeated zone – porewater displaced

Permeation
Low shear; low to very low viscosity grout

TAM *High/low pressure*

Lenses of grout

Fracture
Low shear; moderate to low viscosity grout

Grout tube *High sustained pressure*

Soil compacted

Bulb of grout

Compaction
High shear; high viscosity

High/low pressure

Grout hole

Fissure
Low shear; moderate to low viscosity

Suitable for forming permanent or temporary cut-off barriers by the injection of particulate and chemical grouts under controlled pressure using:

- **permeation grouting** in sands and gravels ('alluvial grouting') beneath earth dams (A L Bell, 1982) or
- **fissure grouting** in rock below dam core trenches and abutments, and around shafts and tunnels.

Effectiveness depends on:

- permeability and size of pores and fissures (determines acceptance and direction of grout flow)
- grout composition, rheology and strength
- method of injecting and the displacement of porewater by the grout without rupturing the soil
- depth to which grout injection pipes can be accurately installed.

Not recommended in aggressive groundwater, which can dissolve grout, or where grout may be eroded under high hydraulic head and groundwater flow.

Ground investigation requires detailed particle size analyses, *in situ* permeability testing of individual strata where practicable, information on water table fluctuations, chemical analysis of groundwater – all to supplement the general conceptual ground model. It is important to retain fines during soil sampling as the groutability of the soil will be reduced by only a small fraction of *in situ* fines.

For a grout curtain in rock the size of fissure, orientation, degree of infilling and permeability in lugeon absorption units (1.3) should be determined.

Design for permeation of porous ground is based on the displacement of porewater by steady, continuous grout injection using filtration principles (Raffle and Greenwood, 1961), at pressures which do not disturb the ground by fracture ('claquage'). Empirical guidance is given in 7.1 on a 'groutability index' for ground improvement which can be applied to grout curtains.

Permeability can be reduced from 10^{-2} to 10^{-6} m/s in medium to coarse soils using best practice, but the treatment is rarely uniform, particularly in finer soils where grout lenses easily form due to soil fracturing.

Incomplete treatment may lead to continuing water flow causing destabilising seepage and leaching of grout.

Fissure grouting can reduce permeability in rock to below 10^{-7} m/s.

Tube à manchette (TAM) is the preferred method for permeation, using 'sleeve' tubes 50 mm in diameter, which are sealed by annular grout into a drilled hole. Grout is then injected under pressure between packers through the rubber sleeve valves over ports at 0.5–1 m intervals down the tube. The 'plastic' annular grout self-heals after each rupturing, thereby avoiding problems with grout break-back and grouting in the packer.

The advantages of TAM:

- ability to target different strata with different grouts
- multiple injections at each port without re-drilling
- more control over volume and pressure.

Disadvantages of TAM (compared with grout lances):

- time to install the system; annular grout setting time
- cost of materials (steel or plastic tubes, etc.)
- limitation of the depth of treatment by the initial hole length
- wear caused by repeated packer inflation
- extended injection times.

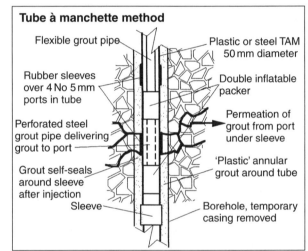

Tube à manchette method

Flexible grout pipe

Rubber sleeves over 4 No 5 mm ports in tube

Perforated steel grout pipe delivering grout to port

Grout self-seals around sleeve after injection

Sleeve

Plastic or steel TAM 50 mm diameter

Double inflatable packer

Permeation of grout from port under sleeve

'Plastic' annular grout around tube

Borehole, temporary casing removed

The grout lance method can be effective for shallow treatment such as underpinning (7.1), but does not have the flexibility for forming a cut-off barrier in soil.

Stage grouting is used for cut-offs in fissured rock, with rubber packers in open drill holes. The stage length and the use of descending or ascending stages (7.5) will depend on the degree of weathering, depth of treatment,

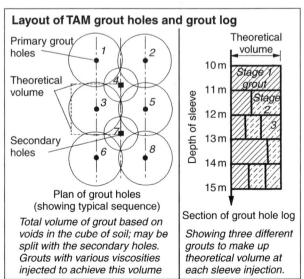

Layout of TAM grout holes and grout log

Primary grout holes

Theoretical volume

Secondary holes

Plan of grout holes
(showing typical sequence)

Total volume of grout based on voids in the cube of soil; may be split with the secondary holes. Grouts with various viscosities injected to achieve this volume

Theoretical volume

10 m
11 m
12 m
13 m
14 m
15 m

Depth of sleeve

Stage 1 grout

Stage 2

3

Section of grout hole log

Showing three different grouts to make up theoretical volume at each sleeve injection.

spacing and size of fissures (Ewert, 1994). For 'contact' or 'blanket' grouting at exposed rockhead, downstage grouting with the packer sealed into concrete is preferred.

Grout hole layout is based on the time for the grout to penetrate the soil, which depends on porosity, permeability and particle shape and the rheology and strength of the grout. Typical spacing is 2–3 m for cement-based grout in gravels and <1.5 m for chemical injection into sand, subject to pressure limitations. Shallow treatment requires closer spacing, <1 m with grout lances. Secondary and tertiary hole location is based on the results of primary injections.

Holes at 1.5–10 m centres, typically in two rows, are used for deep fissure grouting with cement.

The width of the grout curtain is determined by the hydraulic gradient across the cut-off produced by the difference in groundwater levels each side of the curtain – due to either dewatering or impounding behind a dam. A maximum gradient of 5 is suggested for temporary alluvial grout curtains.

The depth of the grout curtain is determined by the depth to the impermeable layer or to where rock is tight enough to prevent significant flow under the hydraulic head; flow nets are needed for partially penetrating cut-offs.

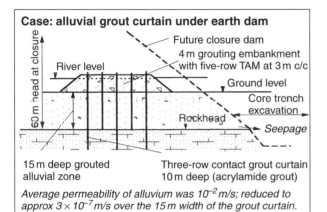

Case: alluvial grout curtain under earth dam

60 m head at closure

River level

Future closure dam
4 m grouting embankment with five-row TAM at 3 m c/c
Ground level
Core trench excavation
Rockhead
Seepage

15 m deep grouted alluvial zone

Three-row contact grout curtain 10 m deep (acrylamide grout)

Average permeability of alluvium was 10^{-2} m/s; reduced to approx 3×10^{-7} m/s over the 15 m width of the grout curtain.

The selection of grout is critical to permeation (7.1 for general properties and applications). Permeation using TAM is designed on the basis of staged injection of different grouts at each injection point:

- stage 1 – relatively viscous particulate grout (say, clay–cement) to fill large voids in gravels
- stage 2 – chemical injection such as one of the silicate formulations
- stage 3 – low-viscosity chemical such as resin-based material.

Clay– (or bentonite–) cement grouts in proportions of 3–5:1 are typical for stage 1 (Littlejohn, 1982). The grouts should have low early shear (gel) strength to ensure that subsequent stages can penetrate to fill remaining voids.

Chemical grouts should have adequate gel strength to avoid erosion under the hydraulic gradient and be stable in varying groundwater conditions (Karol, 1990).

For dam cut-offs in fissured rock, 10–30:1 water:cement grout is normal (Houlsby, 1982). Addition of 2–5 per cent by weight of PFA (pulverised fuel ash) or bentonite will assist penetration and reduce 'filtration' blockages.

Rheology (5.2) of the grouts should be checked to ensure pumpability and reduce potential to separate. The viscosity and setting (or gel) time of the grout govern the initial sleeve lifting pressure and the distance it will travel under pressure from the injection point.

The volume of grout is estimated using pore size. It is assumed that the first-stage grout will fill large pores in a

cylinder of ground around each injection point, with subsequent injections targeting the finer voids.

Where a shortfall in filling of pores occurs compared with the theoretical volume, the balance is made up by injections through adjacent TAMs and secondary holes.

Note that grout is *not* injected until 'refusal' at any one point (i.e. no more grout accepted at that level) in permeation treatment of soils.

In fissure grouting, cement grout is injected until a refusal pressure is reached, either related to the overburden resistance and strength of the rock or as decided by hydro-fracture tests in weak rock.

Jet grouting (7.6) is now favoured for a wider range of cut-offs in soil, producing lower treated permeabilities in overburden comprising fine soils to coarse gravel and weak rock. It is useful for horizontal barriers.

A major benefit is that grout quantities are more predictable.

Construction methods are also covered in 7.1.

Drilling using duplex overburden methods to install casing is required for placing TAMs and the surrounding annular grout. This is possible to depths of 100 m from the ground surface. A grout lance can be driven to around 10 m.

Primary holes for dam grout curtains in rock may be core drilled to supplement geological data. Drilling for the majority of the holes will be by rotary-percussive down-the-hole hammers (11.1), either from the ground surface or from within grouting galleries where depths would otherwise be excessive. Grout hole alignment is critical to optimise the efficacy of the cut-off.

Grout mixing should be carried out in 'high-shear' mixers (11.2) that produce a colloidal suspension of solids; chemical solutions may be mixed in rapid stirrers. The mixed grout should be kept agitated until it is pumped to the grout hole. The grout must be used before it starts to set; bentonite suspensions must be conditioned before being used.

For large projects automatic weigh-batching plants should be used, recording details of every mix; chemicals and mix water are usually dispensed by volume.

Mixing of two fluids in line where short gel times are used is only feasible with accurate dispensing and effective mixing rotors near the point of injection.

Grout injection by TAM requires even, sustained pressure (<1 bar/m depth initially, following lifting of the sleeve and uplift limitations) at low output (5–20 l/min, depending on ground) to achieve permeation without unwanted soil fracture and break-back of grout to the surface.

The length of pumping lines is determined by friction losses and separation of solids in particulate grouts; by gel times for chemical grout.

Monitoring of all materials used in each hole, volume injected and relevant pressure preferably using electronic data-loggers, is essential for controlling permeation – ideally producing 3D imaging of results. Grout density, 'bleeding' (syneresis), viscosity, gel strength, etc. should be checked in an onsite laboratory.

Precise levelling is necessary to check heave during injections; the use of electro-levels on buildings with real-time readouts is now usual in sensitive areas.

For deep curtains, surveying of the drill hole alignment is advisable.

On completion of the treatment, permeability tests should be made in drilled wells throughout the depth of the curtain and further injections undertaken as indicated.

> **Caution**: grout may cause pollution either during injection or by leaching. Environmental restrictions and COSSH Regulations apply to all grouts. The manufacturers' recommendations for storing and handling materials must be followed.

5.2 Bentonite Slurries

Bentonite is used in a wide variety of geotechnical processes – drilling, grouting, supporting trenches and sinking of caissons.

Origin

Sodium bentonite is a naturally occurring colloidal clay composed mainly of the mineral sodium montmorillonite with high specific surface and particles <0.5 μm, found originally in large deposits at Fort Benton, Wyoming.

The chemistry is complex, with a broad, thin platelet-like stacked structure – an octahedral aluminium hydroxyl layer sandwiched between two layers of silicon–oxygen tetrahedra. The sodium ions between the platelets facilitate water penetration, which hydrates the bentonite in a remarkable absorption action which results in its well-known swelling and thixotropic properties.

Bentonite mined in the UK and used in civil engineering works is naturally occurring calcium bentonite (Fuller's Earth) which is treated with sodium carbonate and magnesium hydroxide to produce the sodium form by ion exchange. This 'sodium-activated clay' has the typical swelling and thixotropic properties of bentonite suitable for civil engineering purposes, but the rheological properties of Wyoming bentonite are not fully matched.

Calcium bentonite is used as-mined in Europe generally, requiring higher concentrations than activated clay due to its lower swelling potential. This results in denser and more viscous slurries at higher cost with more wastage.

Properties

The processed bentonite is dried and ground into a fine powder for general use; pellets are made for easier placing in boreholes to seal around *in situ* instruments.

There is no standard UK specification for civil engineering bentonites; the oil drilling industry specification (OCMA Specification, 1973) is based on variations of Wyoming bentonite. The composition and properties of the particular bentonite as provided by the manufacturer must be understood before formulating a slurry for a specific purpose.

Atterberg limits (for consistency) are useful guides to the behaviour of fine-grained materials, bentonite included. Some liquid limit comparisons:

- Wyoming bentonite – 630–750 per cent (depending on additional ion exchange treatment)
- UK sodium-activated bentonite – 400–550 per cent
- London Clay (illite) – 100 per cent
- Boulder clay – 60 per cent

The specific gravity of bentonite is between 2.5 and 2.65.
The pH of Wyoming bentonite is neutral; processed bentonite is around 10.5.
The activity of sodium bentonite is >7 (London Clay is 0.95).

Rheology

The study of the flow of fluids, their viscosity (resistance to flow) and plasticity is critical to understanding the way bentonite slurries work in practice.

Rheological properties depend on the available exchangeable sodium cations, natural Wyoming bentonite being much 'richer' than processed bentonite. When the bentonite is mixed with water a colloidal suspension (slurry) is formed with the charged platelets randomly distributed. When the slurry is left to stand the particles are attracted to each other to form an electro-chemical structure called a 'gel'. When agitated the bonds are broken and the gel reverts to a fluid suspension – the thixotropic process.

Thixotropy is a vital feature of bentonite slurry, and in a true thixotropic fluid the 'sol–gel' process may be repeated many times.

Flocculation will occur when the electro-chemical attraction reacts to other chemicals in the mix – notably cement and electrolytes such as calcium chloride – producing a permanent, non-reversible gel. Aggregated particles may settle out of the slurry, significantly altering the flow properties.

To counteract flocculation of bentonite in saline conditions, Attapulgite may be used to form a thixotropic slurry, but is difficult to control (Jefferis, 1985). Wyoming bentonite will cope with some groundwater salts.

Measurement of the rheological properties of Newtonian fluids (water, true chemical solutions) is relatively simple as the shear stress in the fluid is proportional to the rate of shear applied. Viscous fluids such as paint, bentonite slurry and grouts exhibit plastic flow characteristics with both cohesion and viscosity – these are Bingham fluids.

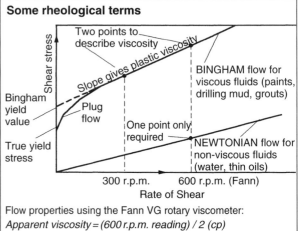

Some rheological terms

Flow properties using the Fann VG rotary viscometer:
Apparent viscosity = (600 r.p.m. reading) / 2 (cp)
Plastic viscosity = (600 r.p.m. – 300 r.p.m.) readings (cp)
Yield value = (300 r.p.m. reading – plastic visocity) × 0.48 N/m²
Gel strength = (3 r.p.m. reading after 10 min standing) × 0.48 N/m²

The rhelogical terms diagram shows the typical Bingham fluid 'consistency' curve produced using a Fann rotary viscometer (Chisholm and Kohen, 1954), where only above a particular rate of shear or rotation is the shear stress increase considered linear. The performance of a range of slurries can be specified and monitored in the field using this curve. It will indicate the 'pumpability' of the slurry or grout, its ability to support the sides of an excavation or borehole and its suitability for injection into soils (Greenwood and Thomson, 1984).

Yield value or stress of suspensions is the value of shear stress required before flow will commence – the measurement of resistance to flow under dynamic conditions. It is dependent on the attraction between particles and can be reduced by dispersants such as sodium polyphosphate without significantly changing viscosity. Flocculated gel will have a higher yield point.

Apparent viscosity is the viscosity at one particular rate of shear and is dependent on the yield stress and plastic viscosity. Two suspensions may have different flow curves but the same plastic viscosity. As water, additives, concentration and solids content all alter the apparent viscosity, variations in apparent viscosity are useful in diagnosing problems in slurry preparation.

Plastic viscosity is the shear stress above yield value which will produce a unit rate of shear. It depends on the concentration, size and shape of suspended solids. The addition of water may be used to lower viscosity, but may only marginally change the yield value.

Gel strength is similar to the yield stress and can be measured with the Fann viscometer after allowing the slurry to stand for 10 min and then shearing the mix until the gel breaks.

Consistency curves and rheology parameters, when used in conjunction with flow and filtration theory, provide the grouting engineer with the basis for the design of grout formulations for permeation, fracture and fissure grouting and can indicate the pumpability and penetrating capability of grouts.

Low-viscosity chemical grouts will perform as Newtonian fluids, but particulate grouts will have complex Bingham characteristics which will affect performance in the soil. For example, the determination of the appropriate rate and pressure of injection from the rheology of the fluid is not only important in achieving success in grouting, but can significantly affect the cost of the work.

When permeating soil, it is important to control the yield stress while ensuring that the concentration of solids will produce the shear strength in the ground necessary to form the cut-off or improve the soil and rock mass strength.

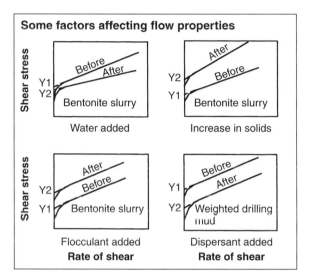

Some factors affecting flow properties

SUPPORT SLURRY

The principle of using bentonite slurry for the construction of slurry trenches and diaphragm walls (5.3 and 5.4) relies on the thixotropic and gel strength properties of bentonite. The slurry will have a greater density than the groundwater, (typically 1.05–1.2 tonne/m³) and by penetrating the surrounding soils will assist in stabilising the side walls of the trench. In addition a 'filter cake' of bentonite gel forms on the trench walls sealing the walls and allowing the full hydrostatic head of the slurry to act. Some filtration of mix water occurs which strengthens the filter cake, but this is counteracted by suspended fine particles from the excavation weakening the filter cake.

To avoid wastage into the soil, penetration of the slurry should be limited (<0.5 m) by increasing the density or by improving hydration before use.

Excavated slurry is screened before being returned to the trench to limit the fines incorporated in the filter cake.

Design is largely empirical, the basic consideration being to ensure that the static head of the slurry is greater than the groundwater head (5.4). The introduction of the lateral soil pressure into the stability equation is complex and it is clear that the arching effect of short lengths of trench wall contributes to stability.

Concentration of sodium bentonite will vary from 4 to 6 per cent for use in soils with permeability from 10^{-3} to 10^{-6} m/s, increasing to 8 per cent in open dry soil, possibly with fillers and additives to increase density and reduce penetration. Essentially, support slurries should:

- be thixotropic and exert sufficient hydrostatic pressure on the sides of the trench or borehole

- form a thin filter cake on the sides of the trench so that the hydrostatic pressure can act
- not easily drain away into the soil around the trench
- not allow solids to settle or water to 'bleed' from the mix
- not be so dense as to affect displacement by the trench backfill, tremied grout or concrete
- be capable of being cleaned of excess solids for re-use.

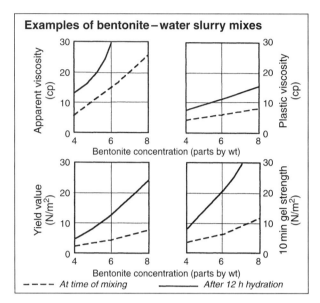

Examples of bentonite–water slurry mixes

Mixing must be done in high-shear colloidal mixers. In high-production plants bulk bentonite is delivered by screw feed with a shaker into a 'jet' mixer. The fresh slurry is transferred to a hydration tank or pond and agitated by circulating pumps for 12–24 h to maximise hydration ('conditioning') before being used as trench support or as an additive to grout.

On large sites it is usual to have a fully equipped rheology laboratory to control the quality and optimise the costs of the slurry. As a minimum the following should be provided for testing to API 13B Specification (1990):

- a Marsh funnel to determine an empirical relative viscosity in seconds for a flow of 1 US quart (water = 26 s)
- mud balance for density
- a graduated measuring cylinder for bleed.

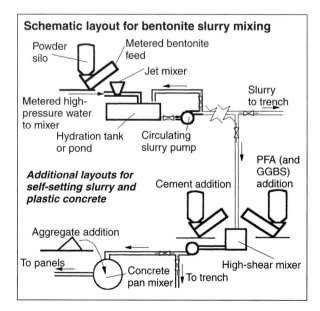

Schematic layout for bentonite slurry mixing

5.3 Slurry Trench Cut-offs

The principle of this cut-off barrier is to form a trench by excavation under bentonite slurry (5.2) to support the sides and then to backfill the trench with either:

- suitably graded gravel to form a soil–bentonite mix or
- a 'self–setting' cement-bentonite grout and a central HDPE or similar geomembrane as the main barrier.

The trench can be 0.3–3 m wide, depending on the design and construction methods used. The base of the trench should toe into an aquiclude for optimum effect, but partially penetrating trench cut-offs may be acceptable in certain conditions (consider implications by using flow nets).

Suitable for in-ground barriers to cut off groundwater flow into excavations, as a means of repairing clay cores in dams and as waste containment barriers to prevent migration of contaminants from new and existing landfill (Privett et al., 1996). Hydraulic permeabilities as low as 10^{-9} m/s are achievable long term, and when combined with a central HDPE membrane, this figure is further reduced. Membrane essential for containment of gases and aggressive chemicals.

Can be installed in most 'soft' ground conditions to considerable depths (100 m without membrane) using diaphragm wall excavators (see 5.4).

Can withstand high hydraulic gradient, subject to the width of the trench backfill.

A slurry trench as defined here is *not* capable of supporting lateral soil pressure as a retaining wall; it relies on support from the ground on both sides and should never be exposed. However, structural support using bored CFA piles to reinforce a self-setting slurry trench (high set compressive strength of 1 N/mm²) has been effective in allowing economic excavation up to a slurry wall (Hayward, 2002).

Case: slurry trench cut-off around core trench of earth dam

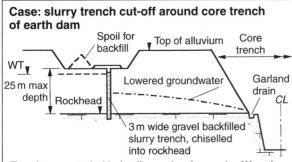

Trench excavated with dragline under slurry using Wyoming bentonite and backfilled with the excavated alluvium.
3 m width of trench designed to withstand 60 m maximum head during closure conditions.
Trench permeability as demonstrated by in situ well tests was an average of 10^{-8} m/s.

Ground investigation should check for boulders which may cause excavation difficulties and trench collapse. Soil strength parameters are required to assess the stability of the trench and bentonite support.

Groundwater levels are critical, as the bentonite slurry in the trench must always produce an excess head of 1–2 m above the water table. It may be cost-effective to reduce or raise ground level in relation to the water table to construct the slurry trench.

Artesian pressure could damage a slurry cut-off; the use of dense plastic concrete backfill may be indicated.

The chemistry of the groundwater and any contaminants will affect the strength, setting capability and thixotropic properties of the slurry (Brice and Woodward, 1984).

Design of slurry trenches requires knowledge of the soil strength and the rheological properties of bentonite and

cement–bentonite mixes to ensure adequate support and long-term stability (Jefferis, 1993).

Specifications for slurry trenches with geomembranes used as barriers to pollution migration are provided in ICE (1999), which is also useful as guidance for design and construction of cut-offs without membranes.

The completed trench should be capable of tolerating moderate deformations without cracking or pulling apart the geomembrane joints.

Self-setting cement–bentonite slurry has the following typical properties, whether used for excavation and permanent barriers or placed following excavation.

Density must be sufficient to provide the excess hydrostatic pressure on the soil – 1.08 to 1.3 tonne/m³ – any higher and there will be difficulty in excavation and placing the membrane.

Density increase due to pick-up of fines during excavation using cement–bentonite slurry is difficult to control – it is not feasible to screen out the fines. In such cases, it is advisable to excavate in short panels rather than as a continuous trench or use the neat bentonite slurry for excavation and displace this with the permanent cement–bentonite mix (as below).

Set strength (UCS) should be a minimum of 50 kN/m² at 28 days, subject to acceptable strain at failure – typically 2 per cent. UCS above 200 kN/m² will increase the possibility of cracking of the set slurry and increased permeability.

The fluid gel strength of a medium-density slurry will be around 20 N/m² to limit penetration into the sides of the trench and initial setting will be evident after 24 h.

Permeability which will meet the minimum strain and strength requirements is unlikely to be $>10^{-8}$ m/s in the short term, say 28 days. Permeability will reduce gradually with time and this must be considered when specifying periods for control testing.

The width of the trench is determined by the hydraulic gradient across the trench – for dewatering cut-offs 10–30, depending on the period of use and chemical nature of the groundwater.

For pollution control a minimum 'permittivity' (= permeability/thickness of trench) of, say, 1×10^{-9}/s has been proposed (ICE, 1999). The contaminant flow rate and migration times also have to be considered.

Apparent viscosity (Marsh funnel) should be around 45 s for pumping slurry up to 1,000 m from the mixer.

Durability data are very limited, but slurry walls have been effective as cut-offs for over 20 years based on piezometer readings on both sides of the wall.

Bleeding (syneresis), when water separates and floats on the surface of the slurry, will lead to instability of the trench; it can be controlled by monitoring mix proportions.

Cement–bentonite mix proportions (by weight) to give a slurry with density 1.106 tonne/m³ are:

Bentonite	50 (CE grade)
OP cement	40
GGBS	70 (to reduce permeability)
PFA	20 (to reduce shrinkage)
Water	1,000

(OP cement is ordinary Portland cement; GGBS is ground granular blast-furnace slag.)

The properties of this mix will generally meet the requirements noted above, but it is necessary to carry out tests for each application to achieve an acceptable compromise between low permeability and deformation and strength.

Where two-stage construction is used (i.e. initial excavation under bentonite which is displaced by the permanent cement–bentonite) the relative densities of the two fluids are critical to ensure displacement.

A geomembrane to reduce the permeability of the barrier is usually 2.5 mm thick HDPE, which has good chemical resistance and hydraulic permeability of 10^{-14} m/s. The composite permeability of the trench will be $<10^{-9}$ m/s, subject to the performance of the HDPE joints.

HDPE sheets of 5 m maximum width with a proprietary factory-welded interlocking joint on each edge are joined as they are slotted into the trench. Site welding of seams is not recommended for this application.

In the absence of a British Standard [prEN 13968 (2002) out for comment], American (American Society for Testing and Materials, ASTM) or German (Deutsches Institut für Normung, DIN) standards are currently used for laboratory tests on a range of geosynthetics (Privett et al., 1996).

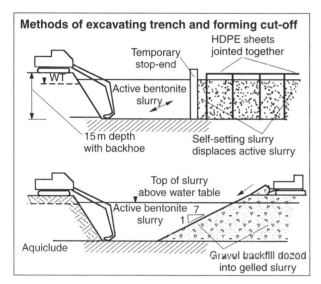

Methods of excavating trench and forming cut-off

Construction of cut-offs backfilled with cement–bentonite mixes is by specialists using either:

- a single stage, in which the trench is excavated under the cement–bentonite slurry, which is then allowed to set as the permanent backfill – satisfactory for shallow trenches where excavation time is much less than setting time or
- two stages, in which the bentonite slurry is used for deeper excavation and then replaced with the permanent cement–bentonite mix.

In both cases the density of the support fluid must allow the placement of a specified HDPE membrane to the base of the trench and avoid flotation of the sheet before setting.

For the two-stage process, the use of temporary stop-ends to separate the bentonite slurry from the permanent mix is desirable to avoid cross-contamination due to:

- suspended solids mixing with the cement–bentonite, reducing permeability and strength or
- cement mixing with the slurry, causing thickening and affecting the potential to re-use the bentonite.

Excavation methods for continuous trenching under the bentonite slurry are backhoe (15 m deep), dragline (25 m) and trenchers (7 m). Excavation for deeper trenches or in uneven areas will require the panel, non-continuous, method of excavation using grabs or reverse-circulation rotary cutters (as used for diaphragm walls).

The trench will generally require toeing into the aquiclude; chiselling into rockhead if necessary. Probe holes, soundings and continuous inspection of cuttings are necessary to define the base of the trench.

In most cases it is advisable to excavate between reinforced concrete guide walls to prevent slumping of the top of the trench due to disturbance of the surface of slurry by excavators.

Mixing of fully hydrated and conditioned bentonite slurry with the cementitious material should be in a high-shear mixer. All solids should be weigh-batched, to ±3 per cent.

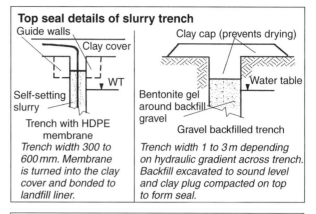

Top seal details of slurry trench

Guide walls / Clay cover / WT / Self-setting slurry

Trench with HDPE membrane
Trench width 300 to 600 mm. Membrane is turned into the clay cover and bonded to landfill liner.

Clay cap (prevents drying) / Water table / Bentonite gel around backfill gravel

Gravel backfilled trench
Trench width 1 to 3 m depending on hydraulic gradient across trench. Backfill excavated to sound level and clay plug compacted on top to form seal.

Caution: disposal of excavation arisings containing bentonite has to be as 'hazardous waste'.

Placing of cement–bentonite slurry to form the permanent backfill should be by tremie pipe so that the excavation slurry is displaced; this may then be cleaned and re-used. Helical rotor displacement pumps are used to deliver the slurry to the trench.

Inserting the membrane requires a roll dispenser or, more usually, a suitable support frame with a sheet attached, lowered carefully into the trench by crane so that it will slide into the joint in the previously placed sheet. Even light winds can hamper this operation.

Backfilling for a wide bentonite–soil trench can be by bulldozing in the graded gravel or placing pre-mixed bentonite–soil with clamshells, with displaced bentonite used for pre-mixing with the gravel.

Capping of the trench is required to prevent drying out and to join the membrane to landfill liners.

Monitoring and quality control of the self-setting slurry (in

Continuous trencher with membrane placing frame

addition to the basic rheological properties of the bentonite slurry; Jefferis, 1993) should include:

- regular samples in 100 mm diameter × 300 mm tubes for UCS and triaxial strength tests and permeability tests
- density by mud balance to check for excess fines
- viscosity using the Marsh funnel
- pH, filter loss and bleeding.

While centralising of the membrane is not usually critical, the self-setting mix should surround it.

Remedial actions to replace sheets damaged during installation or to repair collapsed trench have to wait until the mix has hardened sufficiently to re-excavate.

As for all groundwater cut-offs it is useful to have piezometers on both sides of the wall to check effectiveness.

The in situ permeability of wide, gravel- and soil-backfilled trenches can be checked with tests in bored wells.

5.4 Diaphragm Walls

These are reinforced concrete walls, usually cast *in situ* but may be precast panels, constructed in a trench supported by bentonite slurry (5.2). They are generally part of the permanent load-bearing foundations and perimeter wall.

Suitable for construction of impermeable basement walls (provided joints are adequate) where:

- sheet piles cannot be installed due to obstructions or noise and vibration restrictions
- construction is close to existing structures
- vertical and horizontal loads have to be supported
- simultaneous above- and below-ground construction is needed to meet programme.

They are also used as fully/partially penetrating in-ground cut-off barriers when the trench is filled with plastic concrete. Trench widths are between 600 and 1,500 mm and depth for a cut-off under a dam has been >90 m.
The plan geometry of the wall is not as flexible as that of a secant piled wall (5.6).

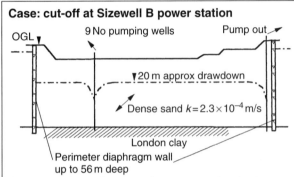

Case: cut-off at Sizewell B power station

Settlement of the adjacent nuclear power station due to dewatering for the new deep foundations had to be less than 6 mm. A plastic concrete diaphragm wall 800 mm wide was therefore used to form a 70,000 m^2 cut-off around the 14 m deep excavation. The permeability of the wall, including panel joints, was limited to 10^{-7} m/s in order to achieve the required drawdown from the pumping wells.

After Howden and Crewley (1995) with permission of Thomas Telford Ltd

Ground investigation for a structural retaining wall must identify the soil parameters that are required for the determination of active and passive lateral pressures which will act on the wall during construction and working life (AGS, 1998).
For a cut-off barrier and embedded structural retaining wall it is essential to determine permeabilities and groundwater conditions – seasonal, artesian – and the potential for boulders and other obstructions to cause instability during excavation. Depth of investigation should be up to 3 times the exposed height of a retaining wall below formation level with particular note of layers of clayey soil and depth to toe-in aquiclude.
Groundwater may need to be lowered so that the support slurry will provide the necessary excess hydrostatic resistance; ground level may need to be lowered so that bentonite is not lost in open dry soil.
Artesian conditions may have to be resisted by a dense excavation slurry.

Design for a retaining wall (3.4) is essentially a structural problem to resist bending moments and shear in the wall produced by lateral pressure on the exposed stem of the wall. Width (600 mm is the usual minimum) and depth are subject to stability requirements either as an embedded cantilever or with props/tie-backs (Puller, 1996) and the need to achieve a cut-off.

Concrete will have a cube strength in excess of 30 N/mm² at 28 days and a density of around 2.4 tonne/m³; the standard slump test will be 150–200 mm.

Design for an in-ground cut-off is concerned with producing a low-permeability, low-strength and cost-effective barrier to groundwater flow, for temporary and/or permanent works. It is different from the basic slurry trench cut-off in that it is usually formed with quality plastic concrete, is much deeper and is constructed by the panel method, as used for structural diaphragm walls.
Its main use has been as a cut-off under earth dams where deformation due to the high vertical loading has to be resisted.
As with the slurry trench, this plastic wall is not designed to be exposed as a retaining wall unless reinforced with piles or other strengthening measures.
Panel length will depend on the excavation method and the strength of the soil. For soils with low angle of shearing resistance and cohesion the panel length must be limited to 2–3 m. In stronger soils length can be 6 m as stability will be assisted by the arching effect of the panel sides. Excavating grabs can cut lengths of 1–2 m.
Width is determined by the hydraulic gradient allowable, depending on the concrete backfill, >30 for a low-permeability plastic concrete.

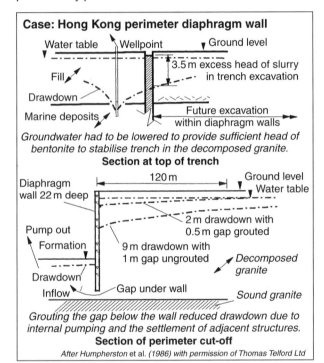

Case: Hong Kong perimeter diaphragm wall

Groundwater had to be lowered to provide sufficient head of bentonite to stabilise trench in the decomposed granite.

Section at top of trench

Grouting the gap below the wall reduced drawdown due to internal pumping and the settlement of adjacent structures.

Section of perimeter cut-off

After Humpherston et al. (1986) with permission of Thomas Telford Ltd

Plastic concrete requirements are typically:

- density of 2 tonne/m³ to ensure bentonite will be displaced
- low bleed but with high water:solids ratio to give low compressive strength, say, 4 N/mm² at 90 days
- minimum shrinkage cracking
- low overall permeability, $<1 \times 10^{-8}$ m/s (including panel joints)
- minimum deformation with specified upper and lower elastic moduli
- resistance to attack by aggressive groundwater.

Mix proportions which generally satisfy the above are typically (by weight):

OP Cement	330
PFA	330
Fine aggregate	2,550
Coarse aggregate	1,650

Bentonite   6 per cent hydrated slurry
Water   1,000
Retarders and plasticizers as needed.

Design of support slurry for retaining wall or cut-off construction should be as noted in 5.2 above for a basic bentonite slurry.

Density will range from 1.05 to 1.2 tonne/m^3, depending on the ground being excavated and the backfill – dense or plastic concrete or less dense sand–cement grout. In order to ensure that the slurry does not pick up excess solids, reducing filter cake effectiveness and making displacement inefficient, the slurry has to be de-sanded (to <3 per cent) using shakers and hydrocyclones before being re-circulated to the trench. The panel construction method means that cement will also contaminate the slurry, causing an increase in pH, loss of thixotropy and flocculation, and treatment for re-use may not be economic.

Apparent viscosity (Marsh funnel) should be around 45 s.

Gel strength of fully hydrated bentonite (4–8 per cent concentration) will range from 10 to 40 N/m^2 before placing in the trench and should be controlled so that filter cake and displacement are not compromised.

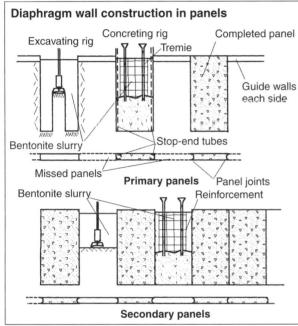

Diaphragm wall construction in panels

Alternative slurries based on high-viscosity polymers are increasingly being used in diaphragm walls and as the flushing/stabilising media for vertical and horizontal drilling. The rheology of polymeric fluids is different from that of bentonite slurry and information on applications to piling and diaphragm walls is still being developed. Benefits claimed for polymers are:

- thick filter cake is not formed on the trench wall
- the lattice structure allows the stabilising hydrostatic force to act and prevents loss into the soil
- pumpability, efficient suspension of excavation cuttings and ease of cleaning for re-use
- possible direct discharge of waste to sewers.

A potential disadvantage in diaphragm walls is that the bond between steel reinforcement and concrete may be reduced due to a residual coating of polymer.
Polymers are initially more expensive than bentonite slurries but repeated re-use makes them cost-effective.
Design is highly specialised and care must be taken when polymer is used in conjunction with bentonite to enhance rheological properties (Beresford *et al.*, 1989).

Construction is usually carried out by specialist contractors. Guidance is provided in ICE Specification (1996a) and BS EN 1538: 2000.
The standard method for both structural and cut-off walls is excavation of primary and secondary panels, with the primary panels achieving acceptable concrete strength prior to excavation of the adjacent secondary panel.

Guide walls are constructed in reinforced concrete each side of the proposed trench to a minimum depth of 1 m – dewatering as necessary. The design must take account of the heavy site loads at the edge of the trench.

Support slurry is run into the trench and constantly maintained at the design level above the water table.

Excavation is by grab hung from a crane or by reverse-circulation rotary cutter (miller) mounted on a crane or purpose-built rig. The rotary cutter has several advantages in that the cutters are interchangeable and the slurry and suspended cuttings are pumped directly to the de-sander units; the clean slurry is returned for re-use. The grab units require muck-away trucks, and loss of slurry can be high.
Verticality is critical for structural walls – usually 1:120 or better; less may be acceptable for in-ground cut-offs provided adjacent panel butt joints are in full contact. The stop-end tube method of forming the joints between panels for structural walls has been improved by various techniques. Special joint formers such as the CWS (continuous water-stop) system provide means of incorporating a water-stop into the joint for panels up to 25 m deep. For plastic cut-off walls, the rotary cutter rigs can cut into the primary panel concrete without the need for a joint former, but contamination of the slurry can be wasteful and the joint may be slightly porous due to slurry residue.
Prior to concreting, the base of the trench should be cleared of debris by air-lift or grab.

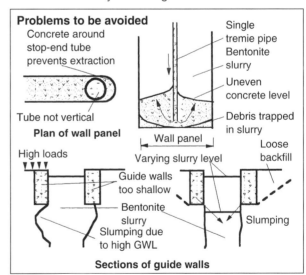

Problems to be avoided

Concreting follows the installation of any necessary reinforcement cage. For structural walls the spacing of the heavy bars must take account of the size of aggregate for quality concrete and the full displacement of the slurry. A high workability is required. One or two tremie pipes are used to feed concrete continuously into the panel at a rate (40–60 m^3/hr) which will ensure the slurry is evenly displaced, the reinforcement is cleared of slurry and concrete flows around the reinforcement.
Plastic concrete will be more workable than structural concrete, but only nominal reinforcement if any is provided.

Monitoring by quality control and quality assurance is usually a strict contract requirement. Slurry rheology and standard concrete properties are measured on a daily basis. Structural behaviour and integrity of the wall are monitored as for bored pile walls (5.6).

5.5 Sheet Pile Walls

Sheet piles are used mainly for temporary ground support to deep excavations and basements where side slopes would be impracticable or uneconomic, and as in-ground cut-off barriers.

Suitable for excavation support as embedded retaining walls and for steel sheet piled cofferdams (a closed cell to exclude water for foundation construction). Depending on soil conditions, exposed heights as retaining walls may be up to 12 m, suitably propped by struts or tied back by ground anchors as excavation proceeds. Maximum driven depth for retaining walls and in-ground cut-offs is 25 m in 'soft' soils.

For groundwater control, piles should toe in to an impermeable stratum, but where this is not possible the precautions for partially penetrating cut-offs must be observed (3.4). Generally, temporary in-ground barriers using sheet piles are not particularly cost-effective compared with slurry trenches as excavation is likely to be easier than driving piles. It can be difficult to achieve relatively impermeable joints – at least 1,000 times less permeable than the surrounding soil is desirable.

Not suitable where noise and vibration restrictions apply or where large cobbles and boulders will damage piles and 'clutches' (the interlocking joints to the sheet piles) during driving or prevent driving to required depth.

Where de-clutching of the piles has occurred due to driving through cobbles, injection grouting or jet grouting behind the piles can considerably improve the watertightness as a cofferdam and cut-off.

Sheet pile walls may be left in place as permanent back-shuttering to the basement or shaft.

Ground investigation should be as for diaphragm walls, with the additional assessment of the driving resistance of the soil based on comprehensive penetration testing of granular soils.

Earth pressures for design will tend to be based on the lower values for the geotechnical parameters provided by the ground investigation, whereas drivability will be limited by the higher values.

Investigation for cofferdams must include determination of groundwater and tidal movements. Also wave, scour and impact forces may be significant for marine sheet piling.

Over-consolidated clay soils may be difficult to penetrate with sheet piles, and as porewater pressures change with time, effective shear strength changes may affect lateral pressures.

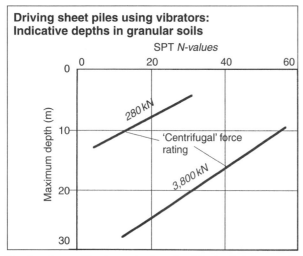

Driving sheet piles using vibrators: Indicative depths in granular soils

Design for cofferdams and embedded retaining walls generally is based on determining resistance to bending and shearing forces imposed by lateral pressure on the wall.

As for other retaining walls (5.4, 5.6), total and effective stress conditions in the soil may need to be considered at different times during construction and service life. Methods of deciding design values of soil parameters based on peak, critical and residual test results for different limit states are given in BS 8002: 1994. Also, empirical rules are set out to allow for:

- softening of clay layers due to water entering tension cracks by applying the 'minimum equivalent fluid pressure' on the side of the wall
- a minimum 'surcharge' load of 10 kN/m² resulting from construction plant and materials at the top of walls.

Determining the size of structural members, props and tie-back anchors requires specialist knowledge; worked examples of cofferdam design are given by Williams and Waite (1993), and Gaba et al. (2003) describe current best practice for embedded walls generally.

Groundwater lowering outside the retaining wall using wells or wellpoints will reduce the lateral pressure and, provided that there is a fail-safe system to ensure that the drawdown is maintained, effective stress parameters may be used.

Once water is removed from a cofferdam, pumping, from either sumps in the formation or wellpoints, will be required to keep the formation dry for construction. Particular attention has to be paid to avoid piping (3.2) and loss of fines at this stage causing instability and danger to personnel in a partially penetrating cofferdam.

Design for in-ground cut-offs assumes that sheet piles can withstand high hydraulic gradients and have low overall permeability – de-clutching or damage during driving will severely compromise effectiveness. Ideally piles should toe in to an aquiclude.

Construction is carried out using a variety of sheet pile sections, piling hammers and methods depending on application, plant availability and environmental factors. Main contractors generally carry out the work in accordance with the ICE Specification (1996a).

Piles are principally 'U'-section, known as the 'Larssen' pile, or 'Z'-section, known as the 'Frodingham' pile. Straight web and 'H'-section piles are also in general use, and many combinations can be made by welding different sections together to improve the bending moment resistance.

Z-section clutches are generally considered more watertight and are preferred for embedded cantilever walls. U-sections are used in pairs to give the specified bending resistance; but with a thicker section than the equivalent Z-section, they are useful in dense and difficult ground.

Pile manufacturers and steel stockholders provide information on the structural properties of the various types currently available.

Butt welding on site to extend pile length must ensure that piles are straight and that the joints are staggered when driven and not at the point of maximum bending moment in the wall.

Hammers for driving piles may be simple drop-hammer types especially in heavy clays, but generally double-acting diesel or hydraulic impact hammers are used for most types of ground.

Medium-frequency vibrating hammers are used where noise restrictions apply (Tomlinson, 1994) and will effectively drive piles in sands, gravels and soft clays. Hydraulic thrusters ('silent' drivers) are useful in clays where noise and vibration would be a problem.

Methods of driving all require a form of guide frame, whether for use as a perimeter wall to an excavation, as a cofferdam or as an in-ground cut-off in order to ensure that the sheets are driven vertically and the clutch connections are as tight as possible. Overhead working space is essential.

The simplest method for short piles in sandy soil is to pitch and drive each single pile to required depth alongside the previous pile; care is needed to prevent twisting and leaning out of vertical. This is generally favoured for high-speed vibrating hammers driving medium-section piles.

The 'panel' method (BS 8004: 1986), controls the verticality and alignment by pitching and interlocking five piles in the guide frame and then driving each to, say, half the required penetration. The next panel is pitched to lock into the last upstanding pair and then driven. The hammer then completes the driving of the first panel and the pitching/driving procedure is repeated for the complete wall. This technique requires a long-jib crane to pitch the piles and care is needed to ensure that the clutches engage smoothly otherwise high friction hinders driving.

The order of driving piles must be carefully planned for cofferdams so that watertight closure at a corner is achieved. Other types of cofferdam and construction methods are detailed in Tomlinson (2001).

> **Caution**: safety procedures must be established and enforced for all forms of pile driving. Safety cages are required for personnel when pitching piles and guiding into the clutch. The pile gates on frames when used as access for personnel must have walkways with safety rails. Shackles for rigging must be inspected regularly.

Driving sheet piles using double-acting hammer
Indicative depths in granular soils

Practical limits for 120 to 150 blows per minute per 250 mm
Minimum section of LARSSEN PILES – Grade 50A steel

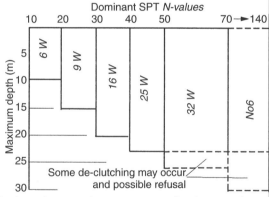

Depths depend on type of strata encountered and construction methods used. It may be necessary to move up a size to achieve the required penetration or greater depth at lower N-value. Note that nomenclature may be different for different manufacturers.

Maintenance of cofferdams is generally concerned with eliminating leaks at clutches which appear at and above formation level after excavation.

In marine conditions leaks can be sealed by dumping PFA in the water outside the sheet piles, which is then drawn into the clutches by gravity flow.

Leaks below cofferdam formation level must not be allowed to cause reduction in the passive resistance of the soil or wash-out of fines. Sealing by injection or jet grouting can be effective in coarse soils. Where leaks are to be controlled by pumping, care must be taken not to remove fines from the soil; internal sumps must have filters, but external dewatering may be indicated as above.

Leaks in an in-ground cut-off wall can be considerably reduced by grouting.

Corrosion of steel is generally not a problem for temporary sheet piles in undisturbed soil, but for long-term use, coating with epoxide paint prior to driving is appropriate. However, where corrosion is likely to cause a severe reduction in the steel sections (marine conditions; hot, humid atmosphere), impressed current or cathodic protection using sacrificial anodes may be needed.

Sheet pile wall with tie-back anchors

Safety procedures for working in cofferdams must include:

- provision of safe working areas, access and alternative exit routes
- life jackets and means of rescue when working adjacent to water
- audible and visual warnings of failure of any part of the system.

Monitoring of the cofferdam must be carried out regularly by experienced personnel and recorded on Form 91 (Part 1 Section B) in accordance with the Factories Act (HMSO, 1988 and TSO, 1996a) to check:

- safe access
- stability of internal struts and frames
- ingress of fines and movement of the sheet piles
- ground movement in and around the cofferdam
- the dewatering system and associated piezometers.

Case: grouted sheet pile cut-off in rock fill bund

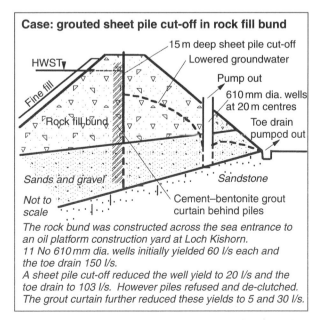

The rock bund was constructed across the sea entrance to an oil platform construction yard at Loch Kishorn.
11 No 610 mm dia. wells initially yielded 60 l/s each and the toe drain 150 l/s.
A sheet pile cut-off reduced the well yield to 20 l/s and the toe drain to 103 l/s. However piles refused and de-clutched. The grout curtain further reduced these yields to 5 and 30 l/s.

5.6 Bored Pile Walls

There are two types of bored pile wall:

- *Contiguous pile walls*, where successive uncon-nected piles are bored in close proximity in a line.
- *Secant pile walls*, where interconnecting piles are formed by first boring and concreting primary piles at centres less than twice the pile diameter and then boring the secondary piles mid-way between and cut-ting a secant out of the primary pile prior to concreting.

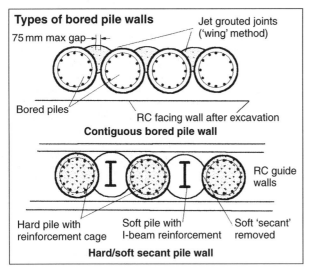

Types of bored pile walls

75 mm max gap

Jet grouted joints ('wing' method)

Bored piles

RC facing wall after excavation

Contiguous bored pile wall

RC guide walls

Hard pile with reinforcement cage

Soft pile with I-beam reinforcement

Soft 'secant' removed

Hard/soft secant pile wall

Suitable for forming walls to basements and groundwa-ter cut-offs where boulders and obstructions would cause problems with driving sheet piles or excavating diaphragm walls. Other advantages are:

- walls with shallow exposures can be constructed close to existing buildings
- length can be varied to suit ground conditions
- large diameters are possible
- low noise and vibration occur during construction
- secant pile joints can be reasonably waterproof
- inclined walls can be constructed.

Suitable ground conditions include most granular and cohesive soils, weak rock and chalk, depending on boring and construction methods – CFA rig, plate auger, percussive boring, etc. with reinforced concrete core.

As with diaphragm walls, 'over-break' will occur where large boulders are removed from the sides of the bore-hole during drilling producing bulges in an exposed basement wall. Also, where weak lenses exist it is possi-ble for concrete to flow from under the temporary casing as it is withdrawn.

Not usually economic for temporary works or in-ground cut-offs. Mainly used as part of the permanent basement wall with an internal lining.

Contiguous pile walls are designed with gaps between the piles and are therefore not best suited for controlling ingress or movement of groundwater unless jet grouting, guniting or shotcrete is used to seal the joints. Hence, these walls are rarely used as cut-off barriers.

Secant pile walls should not have gaps or windows between individual piles – unless there has been deviation of the hole during boring – and are effective where groundwater control is needed.

Ground investigation necessary for using bored piles as retaining walls will include the general requirements noted above for sheet-piled walls and diaphragm walls. Bored piles in loose silts and sands will need to be cased during construction, and any loss of such ground into the pile shaft during extraction of temporary casing will lead to settlement of surrounding ground.

Artesian groundwater should be investigated as piping may occur during concreting.

Design for temporary and permanent earth support will follow the principles outlined for sheet piles and diaphragm walls. The walls may be embedded cantilever, tied back or propped.

The bending resistance of the cross-section of a bored pile means that it is not particularly efficient as an earth-retain-ing structure unless propped. Heavy reinforcement may be necessary, which, in the case of CFA piles, is not usually practicable over 12 m depth. Large diameters (up to 2,000 mm) for cased secant piles can improve bending resist-ance considerably, and tie-back anchors or props reduce the amount of reinforcement needed.

Depth is determined by the structural stability require-ments or depth to an aquiclude.

Spacing of the piles depends on the methods used and the soil and groundwater conditions. For a secant pile wall with 800 mm diameter primary piles in saturated gravel, centres will be around 700 mm. Contiguous piles are spaced so that casings can be installed without interfering with adjacent piles or so that the auger is not damaged by cutting adjacent piles.

Small-diameter (CFA) contiguous piles may be designed and installed in double rows to improve waterproofing, but this is expensive.

Interlocking secant piles usually have reinforcement only in the secondary (or 'male') pile since when cutting into the primary ('female') pile, damage may be caused to the drill and reinforcement. Where necessary for flexural strength, reinforcement in the female piles can be pro-vided by using shaped cages or steel H-beams which will not interfere with boring the male pile.

Concrete strength for contiguous piles using temporary casing is high structural grade (>25 N/mm^2). In the case of CFA piles the strength is dependent on the high slump necessary for concrete injection through the auger stem and the need to push reinforcement into the wet concrete. As the female pile of a secant wall has to be cut by boring the male pile, the female concrete strength can be reduced as in the 'hard–soft' method of construction below.

Shotcrete to piled wall with tie-back anchors

Construction of contiguous and secant pile walls is described in the ICE Specification (1996a). Verticality limits quoted are not as strict as for diaphragm walls – 1 in 75–100. Reinforced concrete guide walls are recom-mended. It is not normal practice to form 'box-outs' or inserts for floor slabs in bored pile walls – standard practice in diaphragm walls – due to difficulty in orienting the reinforcement cage in the borehole.

Contiguous walls constructed using CFA piling rigs (see page 63) (90–150 kNm torque) are quick and relatively cheap to install to depths of 25 m and diameters up to 750 mm in water-bearing granular soils and clay. The insertion of the reinforcement cage into wet concrete at the maximum depth becomes critical – possibly requiring assistance from a vibrator.

Where temporary casing is required to support the borehole or deal with groundwater ingress during drilling (possibly in conjunction with drilling mud), the gap between the piles is increased, and will need to be jet grouted if groundwater is to be controlled when the wall is exposed. Casing must be withdrawn during concreting, and care is needed to keep a head of concrete above the casing toe to avoid 'necking' of the pile in squeezing ground and at the water table.

High-torque auger rigs (300 kNm) are now capable of installing casing to depths of 40 m. Smaller rotary drills can be used to form walls with 300 mm diameter piles in low headroom.

Secant piles to form reasonably waterproof walls and in-ground barriers can be constructed using two methods:

- *Hard–hard,* in which both primary and secondary piles are formed with structural grade concrete (25–40 N/mm^2 at 28 days). This requires a high-torque drill to cut into the primary pile sides to form the interlock with the secondary pile. Even though secondary piles are drilled before the primary concrete reaches significant strength, the use of drill rigs with oscillating casing will ensure that the secondary pile is maintained on line. These walls may be constructed to 40 m depth with 800 mm diameter piles using appropriate rigs.
- *Hard–soft* methods can be used where flexural strength is not critical and for temporary works. Here the 'soft' primary piles are formed with a bentonite–cement grout or low-strength concrete (say, 2–10 N/mm^2), allowing the 'hard' structural concrete (>25 N/mm^2) secondary pile to be formed with an auger cutting into the soft primary (Sherwood *et al.*, 1989).

Standard CFA rigs are used to depths as for contiguous piles in clays and medium-dense gravels.

Piles of 2,000 mm diameter using bored pile rigs can maintain vertical tolerance to better than 1 in 150 to 30 m depth for both types of wall, giving high flexural strength.

Ground anchors (9.1) or props are usually necessary to minimise the volume of reinforcement in the cage as part of the retaining wall design. For the hard–hard method the anchor may be drilled through the pile – subject to reinforcement spacing. For hard–soft walls walings along the front of the wall may be necessary.

Monitoring is required of the integrity of the piles and of the structural behaviour of the retaining wall. The ICE Specification (1996a) advises on the various extensometers, strain gauges and inclinometers to be attached to the reinforcement cage before insertion into the pile bore. The tie-back anchors may also be instrumented (similar requirements are specified for diaphragm walls).

In addition, load cells may be incorporated into the pile. Static load testing is usually specified, together with integrity testing using non-destructive methods such as sonic logging, sonic echo and vibration (Turner, 1997).

Other types of bored walls

The following methods, described in detail in Chapter 7, will provide effective in-ground barriers.

Jet grouting (7.6) can be used to form gravity retaining walls, vertical and horizontal barriers by construction of overlapping, interlocking columns, as in the secant pile method. The barrier is usually made of two staggered lines of columns, with primary and secondary injection holes to maximise interlocking.

The diameter of columns using the 'triple-jet' system in suitable soils is 500–3,000 mm to a depth of 25 m.

Secant pile wall with hard/soft piles

Jet grouting using an oscillating jet in a vertical plane as it is withdrawn can form panels of treated ground which are interconnected.

Good resistance to hydraulic gradient across the wall and low internal permeability ($k = 10^{-9}$ m/s) with either the panel or column method can be achieved using the appropriate cement or cement–bentonite grout.

This method has largely replaced the less sure permeation grouting methods in a wide range of soils.

Soil mixing (7.7) or mix-in-place methods can also be used to form overlapping columns to produce an in-ground barrier. Here the columns are formed by mixing a cement–bentonite grout with the *in situ* soil using a specially designed mixing auger. The wall is formed by a minimum of two rows of staggered augered holes, the column width depending on the diameter of the auger. The soil is mixed either with fluid grout injected through the hollow stem of the auger or with dry reagents. Depth using high-torque auger drives can be up to 50 m. This construction is useful in contaminated soil as it is possible to control dispersal of polluted materials and aerosols.

Contiguous pile wall

5.7 Artificial Ground Freezing

Suitable for forming a temporary cut-off barrier to exclude groundwater in most saturated and near-saturated soils by lowering the ground temperature to form a wall with contiguous ice cylinders. Some soils with moisture content as low as 10 per cent can be frozen satisfactorily.

It is mainly used to assist in the construction of shafts and tunnels, although ice walls have also been used as propped retaining walls around open excavations and as temporary underpinning (J S Harris, 1995).

Depth of treatment will be limited by the accuracy of drilling for the freeze tubes. Refrigeration plant is costly to install but once the ground is frozen the system can be operated cost-effectively for long periods. Liquid nitrogen as the refrigerant is only viable for short-term stabilisation.

Case: freezing in tunnel under East Lancs Road

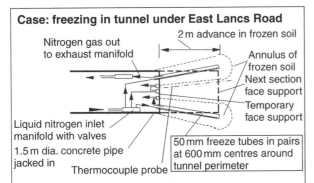

Nitrogen gas out to exhaust manifold

2 m advance in frozen soil

Annulus of frozen soil
Next section face support
Temporary face support

Liquid nitrogen inlet manifold with valves
1.5 m dia. concrete pipe jacked in
Thermocouple probe

50 mm freeze tubes in pairs at 600 mm centres around tunnel perimeter

Pipe jacking 2 m below a dual carriageway had to stop when running sand at the tunnel face threatened major subsidence. Liquid nitrogen was used to freeze an annulus >2 m long which was safely excavated to allow jacking of the next pipe section. Freezing each section took 24 h and the ground remained adequately frozen for 48 h. No heave occurred at the road surface during freezing but the water content of the soil had increased on thawing – no adverse effects were observed.

Case: freezing around shaft in Edinburgh

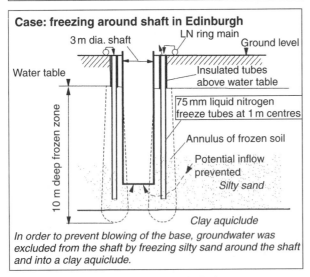

3 m dia. shaft
LN ring main
Ground level

Water table

Insulated tubes above water table

75 mm liquid nitrogen freeze tubes at 1 m centres

10 m deep frozen zone

Annulus of frozen soil

Potential inflow prevented
Silty sand

Clay aquiclude

In order to prevent blowing of the base, groundwater was excluded from the shaft by freezing silty sand around the shaft and into a clay aquiclude.

Ground freezing is not recommended where:

- ground heave may occur during freezing, affecting adjacent structures
- thawing may leave voids in the soil, causing settlement under load and self-weight
- there are flooded cavities or insufficient porewater
- there are variable strata with different thermal conductivities and groundwater flow exists.

The thermal conductivity and heat capacity of the soil and the temperature of the refrigerant used govern the rate of freezing. The strength of the frozen ground depends mainly on the soil lithology, porosity and moisture content, affecting the volume of ice formed.

Ground investigation should therefore determine the following properties as part of the ground model before deciding on the use of artificial ground freezing:

- geological and hydrogeological conditions
- position of the water table and degree of fluctuation
- flow of groundwater
- soil strength parameters
- soil water content and chemical composition
- thermal properties of the soil.

Design has to ensure that the thickness and strength of the frozen ground is adequate to prevent structural failure during construction. An intact cylindrical ice annulus in ground around a shaft or tunnel will be capable of resisting significant hoop stress (Auld and Harris, 1995).

Refrigerants – Brine (sodium or calcium chloride) at a temperature of −30 to −40°C is used if ground freezing is considered at the design stage of a project, such as sinking a deep mine shaft, to lower ground temperature to at least −5°C. A large refrigeration plant is required with reliable pumps circulating brine to the freeze tube system and back to the plant for as long as the ground has to remain frozen.

Liquid nitrogen (LN) at a temperature of −196°C is used as the refrigerant for short-term projects or when emergencies such as unforeseen running sand arise. Only LN is likely to be effective in adequately freezing porewater in cohesive soils. It is fast and efficient in freezing the ground, but as it is vented to the atmosphere during the process and not returned to the plant, it is expensive. The onsite plant is usually a vacuum-insulated pressure vessel with an evaporator to produce flow to the freeze tubes; no power connections are needed. Mobile tankers deliver LN to the pressure vessel from the manufacturer's plant as needed, and once the soil is frozen only a limited amount of LN in the gas phase is needed to maintain the freeze due to the considerable initial sub-cooling adjacent to the tube.

Typical vertical freeze tube for LN

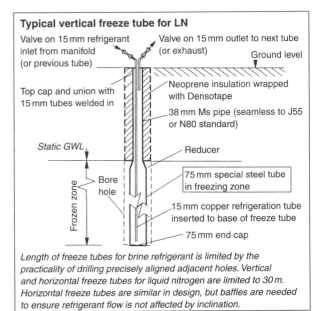

Valve on 15 mm refrigerant inlet from manifold (or previous tube)

Valve on 15 mm outlet to next tube (or exhaust)
Ground level

Top cap and union with 15 mm tubes welded in

Neoprene insulation wrapped with Densotape

38 mm Ms pipe (seamless to J55 or N80 standard)

Static GWL

Reducer

75 mm special steel tube in freezing zone

Bore hole

Frozen zone

15 mm copper refrigeration tube inserted to base of freeze tube

75 mm end cap

Length of freeze tubes for brine refrigerant is limited by the practicality of drilling precisely aligned adjacent holes. Vertical and horizontal freeze tubes for liquid nitrogen are limited to 30 m. Horizontal freeze tubes are similar in design, but baffles are needed to ensure refrigerant flow is not affected by inclination.

Freeze tubes of the type illustrated here at 1 m centres will produce interlocking ice cylinders, with time to freeze depending on the refrigerant. Two rows of tubes to form the barrier are usual for long-term freezing where structural strength and water exclusion are of equal importance. When freezing clayey soil, two rows are desirable to avoid problems with vertical cracks occurring between frozen cylinders. The development of the frozen cylinders is monitored by thermocouples run into probes between a selection of cylinders.

Groundwater flow causes difficulties in producing uniform cylinders around brine freezing tubes: a velocity >3 m/day will require pre-grouting to reduce the flow. Alternatively, dewatering to counter the flow may be effective provided the water table is not lowered at the freezing zone. If LN is used as the refrigerant, groundwater flowing at 30 m/day can be frozen using at least a two-row barrier of tubes. Variations in the thermal properties of the strata to be frozen also cause problems with non-uniform ice cylinders, resulting in gaps in the ice wall. The much lower temperature of LN will mitigate these problems.

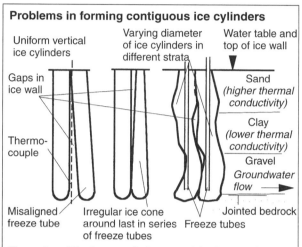

Problems in forming contiguous ice cylinders

Uniform vertical ice cylinders

Varying diameter of ice cylinders in different strata

Water table and top of ice wall

Gaps in ice wall

Thermo-couple

Misaligned freeze tube

Irregular ice cone around last in series of freeze tubes

Freeze tubes

Sand *(higher thermal conductivity)*

Clay *(lower thermal conductivity)*

Gravel *Groundwater flow* ⟶

Jointed bedrock

The radius of the frozen cylinder around the freeze tube will depend on the thermal conductivity, the moisture content of the soil and the refrigerant temperature. In order to achieve an effective ice barrier each adjacent frozen cylinder must be contiguous throughout its length.

Estimated quantities of refrigerant required are based on heat-transfer principles (J S Harris, 1995; Holden, 1997). The graph shows the approximate quantities of LN required for the initial freeze to lower the soil temperature from 10°C to 0°C at a radius of 610 mm (W G Grant, private communication from British Oxygen Company, 1968). Beyond this the growth rate of the ice decreases significantly. The tube diameter only marginally influences the quantity of LN required initially for the same frozen zone, but less LN is used when the tubes are in a liquid/gas series than when all tubes are full of liquid. The total LN requirement is dependent on the time the ground has to be kept frozen.

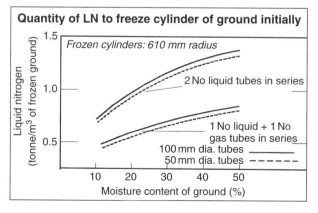

Quantity of LN to freeze cylinder of ground initially

Liquid nitrogen (tonne/m³ of frozen ground)

Frozen cylinders: 610 mm radius

2 No liquid tubes in series

1 No liquid + 1 No gas tubes in series

100 mm dia. tubes ————
50 mm dia. tubes – – – – –

Moisture content of ground (%)

Installation – accurate drilling is a critical part of the process, requiring cased holes, usually by duplex drilling and to considerable depths for shaft sinking, into which the freeze tubes are inserted. It is therefore essential to survey the alignment of each hole so that

additional tubes can be installed where gaps in the ice wall are likely.

It may be necessary to grout around the freeze tubes as the casing is withdrawn to avoid an aquifer discharging into unsaturated strata. If an ice annulus is to be formed around a tunnel prior to excavation, the same care must be taken with the horizontal drilling methods – directional drilling navigation may be appropriate.

Where vertical freeze tubes are necessary to stabilise soil for tunnelling, a technique called 'rolling freeze' has been successfully deployed (Baker and James, 1990). All the vertical freeze tubes are installed initially, but LN is circulated only to the particular section about to be excavated. Note that the steel tubes will cross the line of tunnelling and have to be cut out by miners.

Pipework for freezing deep shafts should provide for separate delivery of brine to each freeze tube off the header main. For LN systems, two or three freeze tubes may be connected in series off the main depending on the depth of the tubes; in this case it is desirable to be able to reverse the flow to ensure that the liquid phase is efficiently used initially to produce uniform cylinders.

For brine and short-term LN freezing it is not usually necessary to fit 'cryogenic' pipework: steel pipe with good tensile properties at −50°C is adequate for the tubes.

All exposed pipes above groundwater should be fully lagged.

Operation – time to freeze soil depends on the thermal properties of the soil, the temperature of the refrigerant and particularly on the diameter of the freeze tube. It will take several weeks to freeze a 1 m diameter soil cylinder using brine in 75 mm tubes, whereas with LN the time will be reduced to around 2 days in similar conditions; increasing to 3 days using LN in a 50 mm tube.

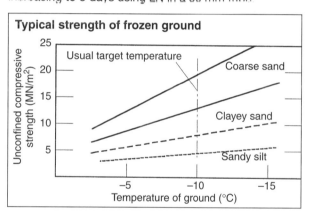

Typical strength of frozen ground

Unconfined compressive strength (MN/m²)

Usual target temperature

Coarse sand

Clayey sand

Sandy silt

Temperature of ground (°C)

Frost heave may occur in some frozen soils, depending on permeability: free-draining soils are less likely to be affected by moisture expansion. In most cases, provided that the confining pressure around the frozen zone is greater than the expansion pressure, heave should not occur. Porewater migration under capillary action during the freezing of silts may increase the volume of water being frozen, leading to increased expansion and potentially unstable soil when the ice thaws. Freezing will permanently alter the structure of peaty soils.

Excavation of frozen soil will require the use of mechanical tools operating in very cold conditions. Construction techniques also have to cope with freezing conditions, but slip-forming of concrete against the ice wall has been successful (Collins and Deacon, 1972).

Caution: during freezing with LN it is necessary to monitor the air quality in shafts and tunnels to detect any accumulation of nitrogen gas leaking from the surrounding freezing system, and to provide ventilation and personnel evacuation procedures (British Cryogenics Council, 1991).

5.8 Groundwater in Tunnels

Exclusion of groundwater from tunnels may be achieved by fissure and permeation grouting and ground freezing to form an annulus around the tunnel. While compressed air, tunnelling machines and sprayed concrete are not strictly geotechnical processes, they are alternative ground stabilisation methods which the engineer and geologist may need to consider when faced with difficult ground conditions in tunnels.

Ground investigation for soft ground tunnelling will need to address the following features to assess suitable stabilisation methods:

- silts and sands above and below the water table – causes of ravelling and running conditions
- overburden pressure and presence of soft/deformable rock, causing squeezing
- over-consolidated clay prone to swelling.

COMPRESSED AIR

Suitable for reducing the volume of water entering confined underground spaces through most types of soil and rock. In soft clays the excess air pressure provides direct support at the face. In silts and sands, porewater is displaced by the compressed air producing some cohesion between particles.

Must not be used where the air pressure required to balance water pressure poses a health hazard, i.e. about 2 bar above atmospheric pressure.

> **Caution**: the monitoring of pressures, air quality, personnel movements, safety, medical examinations, etc. are mandatory.
> The Factories Act (1961), the Health and Safety at Work Act and The Work in Compressed Air Regulations 1996 (HSE, 1996) affect the control of all operations.
> The Medical Research Council/CIRIA *Medical Code of Practice for work in compressed Air* (1982) should be applied.

Because of the health and safety considerations, it is advisable to examine the use of drainage and exclusion methods to control seepage and soil flow into the tunnel before embarking on complex installation and management of a compressed air system. Compressed air tunnels may need some supplementary ground treatment at the shaft such as ground freezing or jet grouting in order to install the tunnelling equipment and air locks.

Ground investigation along the line of a tunnel using boreholes from the surface and appropriate geophysical methods requires careful consideration of potential lateral changes in strata and permeability.

> **Case**: oxidation of pyrite in saturated Woolwich and Reading beds by compressed air resulted in sulphate attack on a concrete tunnel lining when the pressure was released (Eglinton, 1987).

Design is based on balancing the groundwater head by the pressure of air in the tunnel to prevent ingress of water and provide support at the face. Only specialist designers and contractors should undertake this work, as the guidance in the ICE 'Specification for Tunnelling' (2000) and Tomlinson (2001).

Air pressure to balance 1 m head of water is approximately 0.1 bar. In practice, a lower air pressure than the theoretical hydrostatic pressure may be satisfactory in cohesive soil with lenses of water-bearing silts and sands. For a 3 m diameter tunnel below the water table, the head will vary by 0.3 bar between the crown and invert. If the pressure is set at that necessary to exclude water at the invert then there will an excess pressure of 0.3 bar at the crown, which may cause a blow-out. Alternatively, if the pressure is set for the crown then seepage could occur at the invert, requiring internal drainage measures. Experienced site supervisors must decide on altering pressures used to provide safe working conditions.

The volume of air required to maintain the working conditions is determined by the size of the tunnel and the likely air loss into the soil. The safety requirement is 0.3 m³/min of fresh air for each person within the working chamber. Adequate standby compressors must be provided for automatic cut-in in the event of air loss. Loss of air at the face is a costly waste of energy.

Other design details include personnel safety measures, the air-lock systems for personnel and excavated material and venting of excess air.

Dewatering wells may be needed to reduce the water pressure so that the safe working pressure is not exceeded. It is essential that these wells are totally reliable, with standby facilities, as failure of pumping will lead to a rise in groundwater pressure, above the air pressure leading to potential inundation of the tunnel.

Construction under compressed air is arduous, time-consuming and expensive. Personnel working time has to be limited and decompression ('locking-out') time is needed; the spoil has to be taken out through an air lock; air locks have to be moved forward as tunnelling progresses.

Where air is lost at the face and cannot escape to the surface, the head of water may increase, requiring an increase in the air pressure – possibly above the allowable limit. In this case relief wells can both lower the groundwater and safely remove the escaping air until equilibrium at lower pressure is re-established.

Loss of air to the surface may also cause a rise in piezometric head as a result of interaction of permeable soil with the air and water. Fluidised soil in such conditions may flow in from the tunnel face.

Local sealing of the face in coarse soils is necessary in most applications – reliance being placed on the expertise of the shift supervisor. Water seepage at the tunnel invert can be conveniently controlled by inserting well-point 'suckers' for a few metres.

TUNNEL SHIELDS

Hooded shield tunnel boring machines (TBMs) support the roof ahead of excavation in soft ground and allow access to the tunnel face for a variety of cutters and stabilisation work.

Rotating full face TBMs with face plates studded with an arrary of rock cutters, are used in hard sound rock (up to 150 MN/m²) where access to the face is not normally required, although ports may be provided for probing ahead. The diameters of these TBMs can be in excess of 10 m.

A bentonite slurry TBM has a bulkhead located between the unit and the tunnel face to form an enclosed pressure chamber. Bentonite slurry is pumped into the chamber under pressure to balance the soil and groundwater pressure at the face. The cuttings from the face are mixed with the slurry and pumped away to separation units. This method is mainly used in soft ground tunnelling.

The earth pressure balance machine is also a TBM with a bulkhead and pressure chamber. The excavated material itself is maintained at the balancing pressure in the chamber and removed by screw conveyors from the face. Slurry and other additives may be injected into the chamber to assist in the excavation, mixing and removal process.

SPRAYED CONCRETE LININGS

This system is used as initial support for underground excavations in rock and, more recently, soft ground

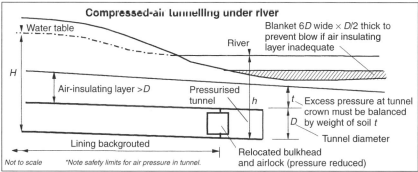

Compressed-air tunnelling under river

Water table

River

Blanket 6*D* wide × *D*/2 thick to prevent blow if air insulating layer inadequate

H

Air-insulating layer >*D*

Pressurised tunnel

h

t — Excess pressure at tunnel crown must be balanced by weight of soil *t*

D

Tunnel diameter

Lining backgrouted

Not to scale *Note safety limits for air pressure in tunnel.*

Relocated bulkhead and airlock (pressure reduced)

---

which requires immediate support. The technique is also known as the New Austrian Tunnelling Method, but care is needed in applying this description to the support of 'soft rock' – clays and chalk (ICE, 1996b).

Sprayed concrete is also referred to as 'shotcrete'; 'gunite' is sprayed sand–cement mortar.

Suitable for primary support in hard fractured rock, possibly in conjunction with steel ribs where high pressure is induced by excavation. Spraying of concrete does not have to follow excavation immediately unless spalling or rock movement is likely.

'Soft' rock (UCS <10 MN/m²), such as weathered chalk, marls and stiff-to-hard over-consolidated clays, requires immediate primary support after excavation. Excavation will be in short lengths ahead of the lining. Use of rapid hardening, dry-fibre-reinforced shotcrete may enable excavation to continue sooner. The permanent support is provided by *in situ* concrete placed against the sprayed concrete lining (Watson, 2003).

Ground investigation for tunnels is briefly outlined above and in 1.4 and 9.5. Engineering geology aspects of NATM are dealt with by Beveridge and Rankin (1995).

Design for tunnels using sprayed concrete as a primary support will initially consider the objectives, structural integrity, durability and costs, as for a tunnel constructed by traditional means (ICE, 2001).

The particular elements relating to the primary sprayed lining in soft ground will require the lining to:

- be placed as soon as practical after excavation and be continuous around the perimeter, with special attention to joints
- prevent stress concentration, loosening and falls.

NATM design and construction relies on the observational method, using an array of monitors for movement at each stage of construction to allow appropriate modifications or remedial actions to be made promptly as the works progress. It is not advisable to alter the design thickness of the lining on site as a result of observations alone – the designer must be involved. Prediction of settlement of surface structures is analysed using finite element methods and may then be controlled by compensation grouting (7.4).

Construction requires careful co-ordination of mining and lining. A variety of methods are used to produce tunnels with primary sprayed concrete linings – pilot tunnel or short headings with benches, or side drifts and temporary linings and supports.

The shotcrete will be applied using either the 'dry' process, where water and accelerators are added at the spray nozzle, or the 'wet' process, using ready-mixed concrete pumped to the nozzle.

Materials include steel sections, wire mesh and fibre for reinforcement, OP cement and well-graded sand–gravel aggregate for the shotcrete. The mix proportions will depend on the process – dry, with low water:cement ratio and high strength but with high rebound; wet for high output (10–20 m³/hr), lower strength, lower rebound.

Plant for onsite mixing requires the use of weigh-batching and accurate dosing of additives, materials handling, pan mixers and pneumatic hand-held spraying equipment or 'jumbo' rigs.

Monitoring relies on a variety of instruments to provide data for the application of the observational method:

- extensometers, precise levelling and pressure cells to monitor deformation and stress in the ground
- strain gauges, load cells and pressure cells for load in the reinforcement and lining.

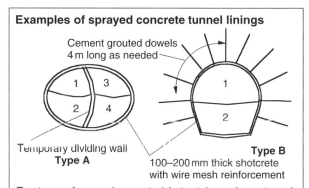

Examples of sprayed concrete tunnel linings

Cement grouted dowels 4 m long as needed

1 3
2 4

Temporary dividing wall
Type A

1
2

Type B
100–200 mm thick shotcrete with wire mesh reinforcement

Two types of sprayed concrete (shotcrete) as primary tunnel lining in soft rock with multiple headings. The sequence of construction as numbered, with heading 1 advancing 2 m ahead of heading 2 and lining following immediately behind. The enlargement in headings 3 and 4 lags 20–25 m behind and temporary dividing wall is broken out.
Dowels are inserted as soon as the shotcrete has hardened.

AUREOLE GROUTING

Refers to grout holes drilled from the tunnel face in a fan or umbrella around the perimeter for up to 10 m ahead of tunnelling. Selected permeation grouts are injected around the tunnel to reduce the groundwater inflow. The heading then continues through the central untreated zone. Jet grouting may also be used to form the aureole. Where faulted zones with high heads of groundwater are encountered, fissure grouting at high injection pressures in the aureole and at the face is necessary. 'Stuffing boxes' to seal around the drill rods and grout pipes are required in these conditions. Temporary sealing of the face may also be necessary.

BACK-GROUTING

In order to fill voids behind the permanent lining, whether concrete or cast iron segments or *in situ* concrete, the annulus between the lining and rock is grouted with cement or sand–cement at low pressure (<5 bar) through ports cast into the segments or the concrete. Later, secondary grouting with neat cement at higher pressure (10–20 bar) and using automatic cut-out controls may be needed to complete the seal if permanent drainage is to be avoided (Henn, 2001).

5.9 Thin Membrane Walls

These in-ground barriers are usually formed by vibrating a steel mandrel with stabilising fins attached into the soil and injecting a plastic grout into the slot formed as the mandrel is withdrawn – known as the 'Vibwall'.

The ETF (a type of 'Vibwall') process uses a heavy steel section driven into the ground and withdrawn to form the void.

Suitable for cut-offs in bunds formed of dredged material around excavations adjacent to the sea and in *in situ* silts and sands. Particularly useful for creating a long length of barrier for temporary works. Maximum depth depends on handling and driving equipment for the steel section – say, 20 m for a vibrated wall.

In the right conditions this is a cheap and effective method for a short-term barrier.

May fully or only partially penetrate the aquifer to be controlled, subject to flow net considerations.

Not suitable for locations where boulders are likely.

As with slurry trenches, this barrier does not have the flexural strength to act as an exposed retaining wall.

Ground investigation should be as for slurry trenches to check depth to aquiclude, presence of boulders and groundwater levels and flows.

Design is essentially concerned with producing a plastic grout which will fill the narrow slot produced by driving and extracting the mandrel/beam, and:

- have low permeability
- not bleed or separate while being injected to cause blockages
- have adequate strength when set to prevent squeezing of the slot.

The grout mix will therefore be based on cement–bentonite slurry thickened with sand (silica flour) and PFA to produce a stiff paste.

The hydraulic resistance of this thin, 100–150 mm wide membrane is limited, hence its use is restricted to temporary works, say, 24–30 months' duration.

Flow nets will be used to determine the seepage under a partially penetrating cut-off and the degree of supplementary dewatering required to stabilise slopes and maintain dry working conditions.

Construction is mainly by the vibrating beam method, using adapted pile vibrators/extractors operating on leaders suspended from a crane to drive the heavy duty H-beam (wide, thick flange beam) into the ground. The beam may be driven as a single unit or as a clutch of two or three depending on the depth required. Overhead working space is required.

The grout pipe is attached to the beam web and slurry or grout may be injected to assist driving. The injection nozzle at the end of the pipe requires frequent replacement. On achieving the designed depth, grout injection starts as the beam is extracted at a rate which matches the grout delivery to ensure the void is filled to form a continuous membrane. It is useful to monitor injection/extraction rates with a data-logger control system.

Deviation of the beam, producing gaps in the wall, is more likely with driving methods than with vibrators; the best alignment is achieved when vibrating a clutch of beams.

The main difficulty with this method is meeting cobbles and boulders or hard ground which cannot be penetrated without damage to the beam and injection nozzle – resulting in a lack of continuity of the wall. It is then necessary to loosen the ground ahead of the beam unit by rotary or auger drilling. This adds considerably to the cost and time.

Monitoring will require piezometers each side of the wall to check drawdown for both fully and partially penetrating cut-offs.

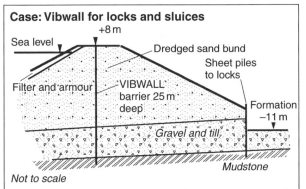

Case: Vibwall for locks and sluices

Sea level +8 m — Dredged sand bund — Sheet piles to locks — Filter and armour — VIBWALL barrier 25 m deep — Formation −11 m — Gravel and till — Mudstone

Not to scale

Because of boulders in the fill and hard lenses in the gravel and glacial till layers, it was necessary to pre-drill on the line of the Vibwall before the steel section could be inserted into the mudstone.

Construction of 'Vibwall'

Vibwall or ETF construction

Grout pipes attached to web — Clutch of H-beams being withdrawn and grout injected — Top of barrier — Guide — Clutch of driven H-beams — Ground level — Completed grouted in-ground membrane — Grout injection

CFA piling rig forming contiguous piled wall

6 GROUND IMPROVEMENT

Ground improvement is undertaken to mitigate settlement induced by changes of stress in the ground, using:

- preventative methods prior to applying the load – e.g. dynamic compaction
- responsive methods to prevent settlement occurring – e.g. compensation grouting.

Shear strength and bearing capacity are also improved. Charles and Watts (2002) and J M Mitchell and Jardine (2002) provide extensive references on treatment and validation.

6.1 Vertical Drainage

The principle of using small-diameter drains in low-permeability soft clays and silts is to accelerate the rate of consolidation under load by draining porewater upwards to a surface drainage blanket or downwards to a permeable stratum. The greater the settlement under the preliminary drainage and pre-load (6.2), the less settlement may be expected when the structure is built. But note that the introduction of these drains into a clayey soil will not reduce the total settlement under the same applied load.

The rate at which a uniform clay layer consolidates depends on the length of the natural drainage path available to dissipate excess porewater – either the thickness of the clay layer or half the thickness of the layer if drainage can be upwards and downwards.

By providing vertical drains the drainage path is reduced to half the distance between the drain centres and access is gained to horizontal partings, laminations or sand lenses in the clay to aid drainage.

Suitable for draining silts and soft clays where horizontal permeability is greater than vertical permeability, but can be successful in stiff glacial tills (Kennard and Reader, 1975).

The process of forming the drain in clayey soil is likely to cause smearing on the sides of the drain hole which will affect performance.

Not recommended in peat and organic clay as the long-term secondary compression (creep) of the soil is likely to be greater than the settlement induced initially by the use of drains (unless combined with vacuum pre-loading below and 4.4).

Other applications include improving the strength of clay under embankments during and after construction, and consolidation of hydraulic fill and plastic clay.

Ground investigation should include continuous sampling of the clayey soil so that detailed inspection and logging of the soil fabric can be made. This will allow selection of appropriate test specimens and targeting of *in situ* permeability tests.

Laboratory oedometer tests (1.2) or *in situ* piezocone and pressuremeters are used to estimate the rate and magnitude of consolidation. As the coefficients of consolidation c_v and to a lesser extent volume compressibility m_v produced from these small-scale tests are not always reliable, it is useful for an experienced engineer to examine the soil fabric and modify the coefficients where appropriate.

Calculations for settlement of clay under load are covered in Tomlinson (2001). Where time is available, the construction of a trial loading embankment with different forms of vertical drains will assist in design.

Design is based on theoretical considerations of permeability, porewater pressure and consolidation, summarised by Hansbo (1993). Procedures for numerical modelling of the consolidation process using drains are now available.

The main requirement is to determine the drain spacing which will give the required consolidation in a specified time. The manufacturers of prefabricated 'band' drains provide design guidance. In addition to the relationship between drain spacing and diameter, the influence of drain transmissivity (well resistance) and the smear zone around the drain is taken into account.

Layout of drains is usually on an equilateral triangular pattern at 1–5 m centres, depending on the type of drain. Ideally, drains should penetrate the full depth of the clay layer. If not, then the consolidation over the bottom 20 per cent of the partially penetrating drain will be slower than estimated, but some improvement will be achieved below the drain tip.

The length of drain required also depends on the ratio of soil permeability to the drain discharge capacity.

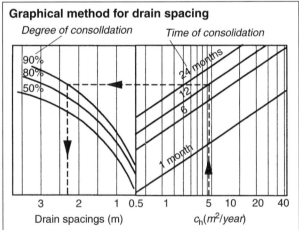

Graphical method for drain spacing

Degree of consolidation　　*Time of consolidation*

This solution is based on Hansbo (1993) for circular drains in a triangular pattern.
For a band drain the equivalent diameter from trials is $d_w = b/2$.
The values given are for a 100 mm band drain, with the smear diameter taken as $2d_w$ and the ratio of horizontal permeability to smear zone permeability as 2.
The drain discharge capacity is estimated from Darcy's law assuming linear flow: $q_w = k_w A_w$ (where k_w is the vertical permeability of the drain and A_w the drain cross-sectional area)

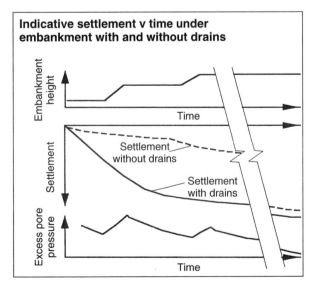

Indicative settlement v time under embankment with and without drains

Yield of the drain can be assessed for sand-filled drains using the Darcy flow assumptions for laminar flow, but manufacturers' data on transmissivity are needed for deep band drains, typically 10^{-5} to 10^{-7} m^3/s for a 100 mm wide drain – higher than for a 300 mm diameter sand drain.

The discharge capacity of drains will reduce with time as consolidation of the soil gradually reduces the horizontal permeability. Also, deterioration and kinking of the band drain material, biofouling and clogging with fines will occur. In most cases the consolidation required is achieved before such deterioration takes effect.

Construction is by pre-drilling holes and filling them with sand ('sand' or 'sandwick' drains) or by driving band drains with a mandrel. The method of installing the drain will have an effect on the smear zone.

A horizontal drainage blanket (300–500 mm thick using rounded gravel filter) to collect the drain discharge is usually placed on the prepared formation prior to installation and will act as the working platform for the rigs; this will have to be cleaned before the structure is built.

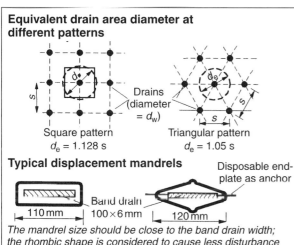

Equivalent drain area diameter at different patterns

Drains (diameter = d_w)

Square pattern
$d_e = 1.128\ s$

Triangular pattern
$d_e = 1.05\ s$

Typical displacement mandrels

Disposable end-plate as anchor

Band drain
110 mm 100×6 mm 120 mm

The mandrel size should be close to the band drain width; the rhombic shape is considered to cause less disturbance to the soil during driving.
The end-plate closes the mandrel and acts as an anchor to hold the drain in place as the mandrel is withdrawn.

Sand drains of 200–500 mm diameter drilled with a continuous flight auger or 300 mm jetted hole are the best means of avoiding significant smearing of the sides of the hole and disturbing the zone around the drain. Spacing is between 2 and 5 m. Maximum depth is 30 m. Driving tubes causes remoulding of the clay and ground heave with closely spaced large drains.

The sharp sand drainage medium is flushed into the hole down a placing tube, or down the hollow stem of a large-diameter auger, ensuring that bridging of the sand in the hole is avoided since this may limit practical depth.

These drains are sufficiently flexible to accommodate some ground movement, but compacted infill sand may act as hard points. Sand drains will operate successfully over a long period if the filter is carefully selected.

Sandwicks are made from geofabric (polypropylene or jute) formed into socks and filled with sharp sand. The 75 mm drain hole is usually formed by driving a mandrel for speed of installation – but this produces some smearing and re-moulding. Drilled holes give better initial drainage. Spacing is from 1 to 3 m, and depth 20 m.

Band drains, prefabricated vertical drains (PVDs), have a core made by specialist manufacturers from thin strips of polypropylene or other synthetic, 3–6 mm thick and 100–300 mm wide. A porous geofabric filter sleeve surrounds the core material.

Hydraulic properties for a 100×6 mm drain depending on core type (BS 6906: 1995) are:

- permeability from 10^{-5} to 10^{-4} m/s
- maximum discharge 0.1 l/s at hydraulic gradient of 1.

Suitable for soft soils with horizontal permeabilities $<10^{-7}$ m/s.

Discharge from a 100 mm wide drain is around 20 m³/year in laminated clay, depending on pre-consolidation pressure and depth, but drains rarely have to perform at this rate. A mandrel driven either by a vibrator or pushed into the soil is used to form a hole, with the drain running off a roll, inserted at the same time. An anchor pin or plate is fitted on the end of the drain so that it is not pulled out of the ground when the mandrel is withdrawn. The upper end of the band drain terminates in the drainage blanket.

The vibrator will penetrate harder soil lenses, but pre-drilling may be required for the mandrel. High porewater pressure can build up while driving and the mandrel can deviate from the vertical – both affecting the consolidation time.

Maximum depth for band drains is around 50 m and outputs of >1,000 m/day are possible.

High settlements can cause deformation of the drain and reduce discharge.

Monitoring of the settlement achieved with time is the most reliable indicator of successful consolidation. It is useful also to record the rate of excess porewater pressure reduction.

VACUUM DRAINAGE

Flow rate from the drains is improved by placing a sealed drainage blanket over the drains and applying vacuum, producing a significant reduction in the time to achieve the required consolidation.

The vacuum has the effect of a pre-load, and is usually applied in conjunction with a loading embankment and membrane over the soil under treatment to form an airtight seal against which the vacuum can act.

The principle of vacuum drainage is to suck porewater out of the clay to accelerate consolidation. It is similar in effect to the vacuum applied to horizontal wellpoints (4.6), but the band drains can be effective to a depth of 50 m, compared with the limited 7 m depth of the wellpoint method.

Suitable for soft saturated clays with permeable horizontal partings; also used for organic silts and clays. Useful in reducing potential problems with secondary settlement.

Construction of the drains is as described above, with a 500 mm deep drainage blanket covering the area. An airtight membrane is placed over the blanket and surcharge fill on top. Vacuum is then applied to the drainage blanket to produce a negative pressure of up to 0.8 bar – equivalent to a sand surcharge about 4 m high.

Where the drainage is upwards to the drainage blanket a wellpoint pump extracts the water from under the membrane. With downward drainage under surcharge the vacuum produced can be higher, but may require deep pumped wells or eductors to remove water from the receiving stratum.

LANDFILL

Soil cleaning using vertical drains under vacuum into a source of clean groundwater below the pollution can lift water through the polluted zone into the drainage blanket and then to a treatment plant for disposal.

Degassing of landfill with passive vertical drains is also effective.

SLOPE DRAINAGE

Vertical drains are also used to reduce pore pressure at the toes of unstable slopes into underlying permeable strata, particularly where there is perched water. Band drains are preferred if the slope is likely to move as they are better able to continue operating if some deformation occurs.

6.2 Pre-loading

The principle is to load the ground with a temporary embankment before the full structural load is applied, in order to cause advance compression and settlement of the ground. When the final load is acting, the ongoing settlement is therefore much reduced or negligible.

Pre-loading is applied up to the value of the structural load and then removed before construction takes place.

Surcharging is a pre-load which is in excess of the final structural load, applied with the intention of producing a greater rate of compression and less rebound on removal.

Both methods are designed to reduce the settlement that will occur after the final load is applied.

Tests have shown that surcharge of loose fill can produce greater pre-construction settlement and lower long-term settlement than dynamic compaction (Burford and Charles, 1992). Time for settlement of loose fill with large air voids and including clay lumps is rapid; hence the surcharge need not be applied for an extended period.

A surcharge load is frequently placed on the drainage blanket over vertical drains (6.1) in compressible clay to increase the rate of consolidation.

Suitable for loose granular fills and soils where, given good drainage ($k > 10^{-4}$ m/s), settlement can be rapid. In soft compressible clays, unless supplementary vertical drains are used to relieve the excess pore pressure induced by surcharging, settlement time will not be significantly reduced. Similarly, where a pre-loading embankment may cause shearing of weak soil, the addition of vertical drains is advisable.

Adequate low-cost material should be available locally for the embankment; alternatively, loading with water tanks/bags may be feasible using natural sources.

Not recommended where time is limited for construction, unless a staged operation is possible (Tomlinson and Wilson, 1973).

Short-term pre-loading of peaty soils will not be effective if the secondary settlement is likely to dominate.

In cases where buildings or machinery or embankments are to be built on compressible soil or fill but cannot tolerate some ongoing settlement, the use of piles or a piled raft is likely to be preferred by designers (Card and Carter, 1995).

Applications are mainly to improve the shear strength and control total and differential settlement of fills, but some landfills are difficult to treat. The method is useful where embankments join viaducts or other more rigid structures.

Ground investigation must determine the strength characteristics of the soil (and of fill if this is to be treated and remain as a foundation material) so that detailed settlement calculations can be made for each layer of soil. Permeability and consolidation parameters are required, together with elastic modulus.

The settlement of foundations on fill will occur due to:

- its own weight and degradation of landfill
- applied load
- consolidation of the underlying ground.

Large-plate loading tests using skips or the construction of test embankments are recommended as part of the investigation to check whether the rate of settlement under pre-load will be effective.

Design requires assessment of the three phases of settlement – immediate, consolidation and creep (Tomlinson, 2001) – with and without the effects of the pre-load.

In weak soil the temporary loading embankment will have to be built up in short lifts so that shear failure does not occur. Vertical drains may improve the subsoil, allowing greater lifts and final height.

As a guide, the ratio of maximum effective vertical stress during surcharge to that due to the structure will be around 1.5. However, to minimise rebound of the ground on removal of the pre-load, surcharge greater than 3 times the imposed load may be needed.

Where a layer of permeable sand exists below the compressible clay the drainage under surcharge will be downwards. The water squeezed out can then be extracted from the sand with pumped wells – this can be accelerated by vertical drains into the sand layer. Where the drainage is mainly upwards a drainage blanket, preferably with additional applied vacuum under a membrane cover, and collector drains are required around the pre-load fill.

Construction area of the surcharge must be larger than that loaded by the permanent structure. The amount of pre-load depends on the consolidation period and the rate of placement.

The pre-load/surcharge is removed when the degree of consolidation is at least 75 per cent of the total estimated; but if time is available it is worth waiting until 95 per cent is reached.

Monitoring will check the rate of settlement and pore pressure reduction. On weak foundations measurement of lateral movement at the toe of embankments may be appropriate.

Band drain layout

CPTs and triaxial tests before and after loading are routinely performed.

Cost comparisons of ground improvement methods ranging from pre-loading to piling:

- drainage and pre-loading are the cheapest per square metre to apply, but the time period required to produce improvement can be considerable

Band drain rig

- dynamic and vibro-compaction are next in cost and results are virtually immediate
- jet grouting and deep soil mixing are 2–2.5 times the cost of drainage/pre-load, but provide considerably greater strength improvement
- piling and vibro-concrete columns for areal treatment are 4–5 times the cost of drainage/pre-load.

6.3 Vibro-compaction

Also known as 'vibro-flotation'; refers to deep compaction of soil with a vibrating 'poker' (or 'vibro-flot').

The principle is to increase the density and load-carrying capacity of loose partially and fully saturated soil by the vibration and displacement of the particles by the vibrator. The improved inter-granular friction reduces settlement under applied load.

The vibrator is either flushed down to the required depth in the soil to be treated using water jets or vibrated dry with air jets in partially saturated soil. As the poker is withdrawn the horizontal vibrations cause a compact cylinder of soil to be formed at depth with a depression at the surface. This depression has to be filled with imported granular material or with adjacent site material and vibrated.

The treatment is permanent and not affected by groundwater conditions.

Suitable for increasing the density of loose sands (zone B in the diagram); other soils can be treated, but greater energy and time are required outside the preferred grading zone. It is not effective in soils with high silt content (>20 per cent); but stone columns using vibro-replacement (6.4) are used in these conditions.

Not recommended where there are obstructions, cobbles or boulders which will damage the poker. Loose fill, particularly landfill, is not amenable to vibro-compaction.

Applications in addition to reducing settlement include improving resistance to earthquake-induced liquefaction; making flexible pavement construction or low-cost warehouse slabs possible directly on compacted soil; reduction of coarse soil permeability (by 2–3 orders of magnitude) to facilitate dewatering.

Hydraulic fill has also been compacted by vibration.

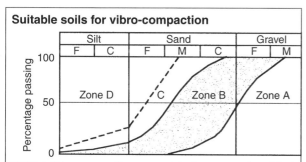

Suitable soils for vibro-compaction

ZONE A: Vibro-compaction appropriate, but penetration difficult.
ZONE B: Most suitable for vibro-compaction.
ZONE C: Vibro-compaction feasible, but longer time required.
ZONE D: Vibro-compaction not feasible – use stone columns.

Ground investigation must determine the particle size distribution (avoid loss of fines), void ratio, soil stratification, groundwater conditions and relative density.

Design is empirical and based on engineering judgement depending on equipment and procedures as well as soil conditions. The most effective design is obtained from the degree of improvement achieved in site-specific trials. The ICE Specification (1987) gives guidance. Concrete footings must be reinforced.

Layout of compaction points is based on a radius of compaction of around 1.5 m in granular soil and a grid pattern selected to suit loading requirements:

- rows of compaction points under strip foundations at 1–2 m centres
- overlapping 2–3 m triangular grid for areal treatment

Depth of compaction can be 35 m in special cases, but 10–15 m is typical. Compaction not likely in top 1 m.

The strength of treated ground will provide allowable bearing capacity of 100 kN/m^2 to over 400 kN/m^2 on well-compacted sands.

The target relative density after treatment is 75–80 per cent.

Construction in granular soils requires the use of high-pressure water jets from the tip of the vibrator to flush out fine soil and to assist in forming the hole as the vibrating poker is slowly lowered under its own weight. When the poker reaches the required depth it is surged up and down to remove fines. The water flow can then be switched to upper jets at lower pressure and the granular fill (20–50 mm grading) added as the vibrator is withdrawn in 0.3 m stages up to ground level.

With electric vibrators the operator will see a specified increase in the current to the vibrator (through a simple ammeter recorder/readout in the cab) before moving up to the next stage.

Outputs range from 30 to 90 m of treatment per hour.

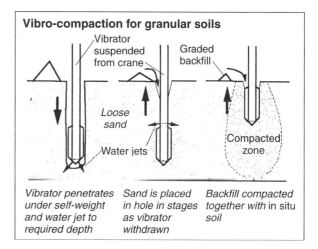

Vibro-compaction for granular soils

| Vibrator penetrates under self-weight and water jet to required depth | Sand is placed in hole in stages as vibrator withdrawn | Backfill compacted together with in situ soil |

Vibrator equipment is either electrically or hydraulically driven and suspended from a crane for operation. The power rating for vibrators ranges between 100 and 150 kW, with eccentric impact force of 300 kN and amplitude of 20 mm. The poker is 3–5 m long by 350–450 mm in diameter with stabilising fins, and is separated from extension tubes by a vibration isolator.

Water and air for jetting are fed into the vibrator ports through flexible hoses. For routine compaction to 10 m deep a 20 tonne crane with suitable jib is required; 20 m treatment depth requires a 60 tonne crane; 35 m depth is possible with extension tubes.

Monitoring is by comparing SPTs or CPTs before and after treatment. Simple non-standard dynamic probes can be equally effective in comparing values before and after improvement – but should not be compared with the SPT or CPT.

In order to assess bearing capacity and settlement potential, zone or large plate loading tests are required.

Vibrator and cone testing probe

6.4 Vibro-replacement – Stone Columns

The vibro-replacement technique uses the vibrator or poker to displace and compact the *in situ* ground, as in vibro-compaction. But in this case, rather than collapsing the soil, a hole is formed into which stone is placed and compacted to form a load-bearing 'stone column'.
Expert assessment is essential.

The principle is to replace loose material by compacted stone, together with densification and reduction in compressibility of the surrounding ground to form a composite material. Because the stone columns will deform under applied load, the capacity of the columns depends on the degree of stiffening achieved in the surrounding soil as well as on the internal friction of the columns.

In fine soils the stone columns will act as drainage paths to accelerate the rate of consolidation – provided that the drained water can be dispersed.

Suitable for reinforcing most loose granular soils, fills, weak and cohesive soils; shear strength of the soil to be treated should be >20 kN/m².
Not recommended where peat layers exist, nor in loose collapsible fills with voids. Obstructions to penetration should be avoided – e.g. rubble in demolished basements. In loose unsaturated fills stone columns may provide a pathway for surface water into underlying untreated soil, possibly causing settlement.

Applications include foundations for low-rise structures, strip footings and light rafts, venting wells on methane-generating landfill, reduction in the liquefaction potential of fine soils (Priebe, 1998) and embankment foundations.

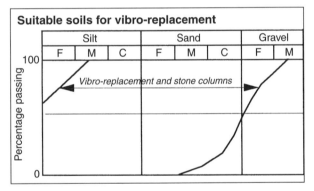

Suitable soils for vibro-replacement

Ground investigation is as required for vibro-compaction (6.3). The chemical composition and levels of contamination of the ground to be treated are important when considering the type of stone to be used for the columns. Penetration tests to determine the initial density of the ground are important.
On complex and large sites a trial treatment with large-scale loading is recommended.

Design method (mainly empirical) will depend on the ground being treated as the interaction of the soil, stone column and foundation is complex.
Ultimate bearing capacity of a single stone column in cohesive soil can be obtained using the Hughes and Withers (1974) method:

$$q_{ult} = \frac{(1 + \sin \phi)}{(1 - \sin \phi)}(\gamma z + 4c_u + q - u),$$

where ϕ is the internal angle of friction, c_u undrained shear strength of the clay, γ the effective unit weight, z the depth, q any surcharge and u the pore pressure.
The ratio of column length to diameter is critical to bearing capacity and should be <6 in cohesive soil.
Settlement of a single stone column can be estimated from Bauman and Bauer (1974). However, because of the variation in the actual size of columns compared with the assumptions, neither of the above mathematical approaches is particularly reliable.

Simple, but conservative, design charts (Moseley and Priebe, 1993), expanded by Priebe (1995), consider load distribution and lateral support from the combined stone column–stiffened surrounding ground on an area basis to give an 'improvement factor' n. The improvement factor indicates the increase in compression modulus and the extent to which the settlement will be reduced by the column/ ground improvement. This is compared with the area ratio (the ratio of the area being treated by each column A – based on the column centres – over the stone column plan area A_c). In soft compressible soils, the stone columns are likely to bulge and settle and transfer applied load to the clay; if no bulging occurs the column carries all the load on to the bearing layer.
The BRE Specification (Watts, 2000) also provides information on design and construction.

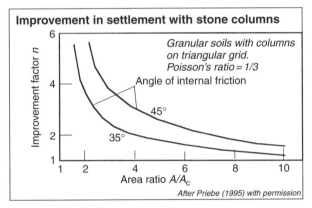

Improvement in settlement with stone columns

Granular soils with columns on triangular grid. Poisson's ratio = 1/3
Angle of internal friction

After Priebe (1995) with permission

Depth of treatment can be obtained directly from the Priebe graphs. Where possible the stone columns should penetrate fully to a sound bearing layer. The type of rig and placement method for the stone will limit the depth possible – around 10 m for bottom-feed rigs.
The top 1 m of the treated ground will be less well compacted and will require some additional rolling before placing shallow foundations.
Long stone columns under concentrated loads are not advisable as the load is dissipated into the ground at a relatively high level as the column bulges.
Layout of columns will depend on the foundation being supported:

- a single line of stone columns under strip footings
- a triangular grid for areal treatment – at around 1–3 m centres depending on the diameter of column which can be formed and subject to remaining within the Priebe (1995) parameters.

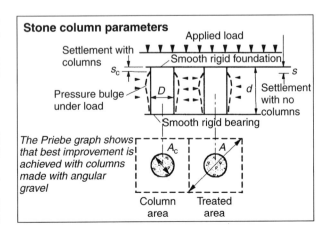

Stone column parameters

Applied load
Settlement with columns
Smooth rigid foundation
s_c
s Settlement with no columns
Pressure bulge under load
D
d
Smooth rigid bearing

The Priebe graph shows that best improvement is achieved with columns made with angular gravel

A_c | A

Column area | Treated area

The strength of treated ground gives allowable bearing pressures for foundation design of 150–400 kN/m² depending on the type of ground treated, column spacing and whether the columns are fully or partially penetrating to a bearing layer.

The long-term performance of the stone column may be affected by deterioration of the stone aggregate and reduction in the support of the surrounding ground.

Construction uses similar plant to that for vibro-compaction (6.3), but here the stone used is inert, crushed, graded aggregate normally in the range 40–100 mm. The stone used must be selected so that it will remain stable for the life of the structure being supported – e.g. limestone is not suitable for acidic conditions.

If hard layers exist in the ground to be treated, it may be necessary to pre-drill before inserting the poker to form the columns. The benefit of speed of installation of stone columns is then lost.

The following are methods for stone columns using vibro-replacement.

The top-feed system is used in dry conditions in stiff soils, where the poker penetrates the ground under its own weight using compressed air and vibration. At the required depth the poker is withdrawn and a small charge of the stone is introduced and compacted by the vibrating poker to interlock with the ground. The process is repeated by adding and compacting stone until a dense column of stone is built up to ground level.

The diameter of columns will be limited to the poker diameter in stiff soil.

The wet system is used with top feed of the stone in cohesionless soils below the water table and in weak silts and clays. The vibrating poker penetrates to depth under its own weight, with water jets removing fines. At the required depth the water pressure is reduced to ensure that the hole formed remains open with water circulation and stone is introduced down the annulus and compacted in short lifts to build up the column to the surface. Column diameter will usually be larger than the poker.

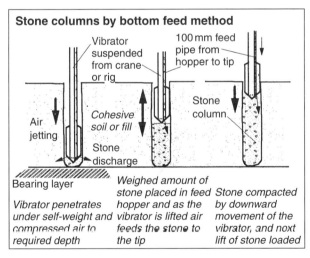

Stone columns by bottom feed method

Vibrator suspended from crane or rig

100 mm feed pipe from hopper to tip

Air jetting

Cohesive soil or fill

Stone discharge

Stone column

Bearing layer

Vibrator penetrates under self-weight and compressed air to required depth

Weighed amount of stone placed in feed hopper and as the vibrator is lifted air feeds the stone to the tip

Stone compacted by downward movement of the vibrator, and next lift of stone loaded

The bottom-feed system can be used below the water table without water jetting, thereby avoiding the problems of disposal of the surplus water. In addition to penetration under its self-weight, compressed air and vibration, the rig may provide additional pull-down force on the poker. The poker has a stone supply tube and hopper attached to deliver stone to the tip of the poker assisted by compressed air. At the required depth the poker is withdrawn 1 m, and a designed amount of stone is discharged at the tip and then compacted by the surging action of the vibrator. The process is continued until the compacted stone column reaches the surface.

Some ground heave may occur in cohesive soils.

Plant and equipment required is similar to that for vibro-compaction: a 300–450 mm diameter poker weighing 3–5 tonnes. Specialist rigs are now common for the dry bottom-feed system.

Cased and rammed stone columns are used in high-sensitivity clays where vibro-techniques are likely to weaken the surrounding soil (Chummar, 1998).

Monitoring using CPTs or SPTs has limited value in verifying the dimensions and internal angle of friction of the column. An improvement in *N-value* of 10 should be achieved in the surrounding granular soil. Stiffening is unlikely in cohesive soil, causing 'hard points' at the columns.

Zone loading tests over a group of columns are the best means of ensuring the settlement and bearing capacity. Construction quality is monitored by the power take-up of the poker and the rate of stone placement.

VIBRO-CONCRETE COLUMNS

Here, concrete replaces the stone as the column infill medium to provide higher load capacity and settlement control where the soil is unsuitable for stone columns – e.g. compressible peat and clay (Maddison *et al.*, 1996).

Suitable for supporting specific point loads or as a deep support element for pavements and warehousing using geotextile reinforced granular fill slabs on weak soils.

Useful on contaminated sites since spoil is not removed.

Layout is on a strip or triangular grid basis at 1–3 m centres for a maximum depth of around 10 m, depending on the rig being used.

Construction is similar to that used for bottom-feed dry stone columns. The vibrating poker with a tremie tube attached penetrates to the required depth, compacting the surrounding soil. At the bearing stratum the poker is withdrawn 1 m as concrete is injected and then lowered again to compact an enlarged end bulb. Concreting continues as the poker is withdrawn at a rate compatible with the pumping rate until the hole is full of concrete.

Outputs can be 400 m of 450 mm diameter column per day.

Concrete strength is between 20 and 25 N/mm² with a slump of 70 mm.

Typical working loads on a vibrated concrete column of 450 mm nominal diameter are 100–400 kN, depending on ground and bearing conditions.

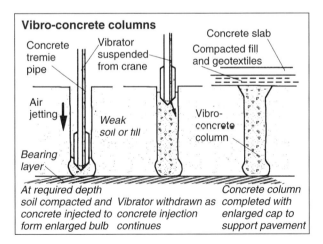

Vibro-concrete columns

Concrete tremie pipe

Vibrator suspended from crane

Concrete slab

Compacted fill and geotextiles

Air jetting

Weak soil or fill

Vibro-concrete column

Bearing layer

At required depth soil compacted and concrete injected to form enlarged bulb

Vibrator withdrawn as concrete injection continues

Concrete column completed with enlarged cap to support pavement

Monitoring will include a continual readout of the poker current, penetration and withdrawal rates and depths plus concrete injection pressure and flow, displayed on a data screen in the rig cab.

Loading tests, as for piles on single columns and zone loading tests for pavements, may also be specified.

6.5 Dynamic Compaction

Widely used for the *in situ* densification or consolidation of granular soils and fills.

The principle is to drop a weight ('tamper') from a height on to the soil surface to compact the ground.

Repeated impacts reduce voids and induce displacement of the soil. Once voids have been closed additional impacts do not improve compaction. Pore pressure must be allowed to dissipate.

Suitable for compaction and improvement of bearing capacity of saturated free-draining soils ($D_{50} > 0.1$ mm) and loose and partially saturated fills where the response can be immediate.

In clays ($D_{10} > 0.06$ mm) the tamper impact causes hydraulic fractures and squeezes out porewater; the response takes time – hours or days after each impact.

Not applicable in soft clay (undrained shear strength <30 kN/m^2), where soil fabric can be compromised by remoulding.

If the water table is within 2 m of the starting level, effectiveness will be reduced – dewatering may be advisable.

The method results in relatively low airborne release of landfill contaminants.

It is economic for sites $>5,000$ m^2; 60 m clearance is needed from other structures.

Applications include compaction of landfill (check anaerobic bacteria action and unpredictable settlements), hydraulic fill and general fills.

Ground investigation should provide information on particle size, voids, groundwater levels and relative density from CPTs and SPTs. It is important to determine the presence of hard lenses within the depth of treatment. Where fill materials are to be compacted, trial pits should be logged in detail and the history of tipping examined. Where soils are assessed as marginal for dynamic compaction, treatment of a trial area is recommended.

Design methods are empirical for spacing of imprint craters and depth of treatment. The ICE Specification for ground treatment (1987) and BRE 458 (2003) provide general guidance.

Induced settlement in natural clay is 1–3 per cent of the compressible depth and in granular soils and fill, 5–15 per cent.

Depth of treatment required will be related to the stress distribution from the foundation load. Compaction down to 12 m is possible in suitable granular soil. Consolidation of cohesive soils and clay fills will be limited to 5 m.

Depth of treatment by dynamic compaction

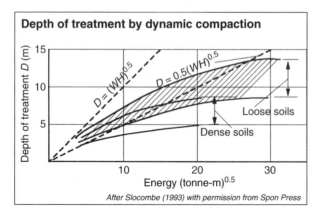

After Slocombe (1993) with permission from Spon Press

Maximum depth can be obtained from $D = n\sqrt{(WH)}$, where W is the weight of tamper and H the height of drop; n varies from 0.4 to 0.8 depending on soil type and speed of drop. Alternatively (more conservatively) $D = WHK/B$, where K is a factor based on soil resistance (between 0.1 for clay and 0.16 for granular soil) and B is the base area of the tamper.

Speed of impact of the tamper should be between 15 and 25 m/s to achieve depths estimated as above, the lower speed being applicable to clayey soils.

The total energy input (number of drops $\times$ drop energy/area) is typically 200–400 tonne-m/m^2.

Typical drop pattern for areal treatment

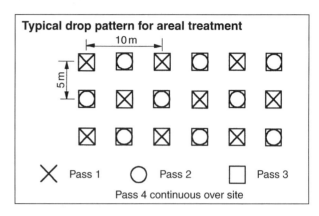

The layout of imprints is designed to treat the deepest layers first with the maximum energy available, on a wide grid with a suitable number of drops at each imprint. The next series of imprints are made between the initial craters with reduced energy, and finally the surface layer is treated with a continuous pattern of low-height drops.

A 'raft' of treated ground at the surface 3–4 m deep will be needed for foundations for storage tanks to reduce the potential for differential settlement.

Treatment depth for low-rise structures is usually not greater than 5 m and the impact passes will be concentrated under the strip footings.

Clayey soils require a greater number of smaller load tamping passes ('mini-treatment') to avoid adjacent heave.

The strength of treated ground will be greatest within the top half of the depth of treatment. Allowable bearing capacity on clayey soils or fills can be 100–200 kN/m^2 and on granular soil up to 500 kN/m^2.

Construction using tampers of 10–20 tonnes and drops of up to 20 m from crawler cranes is typical; special rigs capable of dropping over 100 tonnes from 25 m have been built but are rarely used. The tampers are steel boxes (say, 5–10 m^2) filled with concrete; the shape of tamper (say, a conical base) seems to have little effect on the depth of compaction.

Dynamic compaction sequence

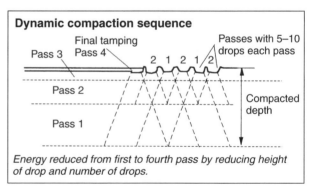

Energy reduced from first to fourth pass by reducing height of drop and number of drops.

For consolidation of clay fills with a high water table or soft clay surface deposits it is necessary to place a free-draining blanket 500 mm deep over the area to be treated and provide drainage channels with environmentally safe means of disposal of groundwater. Pounding wet surfaces can cause soil to move sideways if high-energy drops are used for the first pass.

Initial imprints (first pass) are made on a 5–10 m grid for large areas and directly under strip footings at 2–3 m centres. The number of drops at each imprint (5–10) depends on the soil and consolidation aimed for; craters will be 0.5–2 m deep.

The imprint is backfilled (with free-draining material if necessary) and the second-pass imprints are made between the initial craters with 5–10 drops each (but see 'Dynamic replacement' below). Finally a continuous tamping ('ironing') is performed over the backfilled imprints, using height and number of drops as required from the continuous monitoring.

> **Caution**: A safe distance must be maintained where more than one unit is working on the site. The energy in the loose, falling crane rope can be considerable for high-energy drops – better to use a hook release mechanism.

Monitoring has to recognise the time-dependent nature of consolidation; several days may be needed before the next pass will produce improvement. Dry clay fill will respond more rapidly.

The number of drops, weight and height of drop used and volume of backfill should be recorded.

Ground response – heave and imprint shape (one test for 4,000 m²) – will indicate to experienced personnel the need to modify the energy input to achieve the desired depth of treatment.

SPTs and CPTs are commonly used in granular soils to show ongoing improvement in density between imprints during construction and on completion to compare with pre-treatment values. *N-values* can be doubled.

The pressuremeter was specifically developed for testing compaction.

Zone loading tests, say one every 5,000 m², are usually specified to compare settlement, differential settlement and bearing capacity with required foundation performance.

> ### Case: dynamic compaction reduces liquefaction potential in the Dominican Republic
> A large tourist development was planned on a 34 Ha site subject to liquefaction caused by infrequent earthquakes. The soil was mainly silty sand and clayey silt down to 16 m, which could be densified to over 70 per cent relative density and thereby avoid the liquefaction dangers from 0.2 g acceleration.
> Forty tonne weights dropped 35 m from tripod rigs to pound the site improved relative density to 79 per cent. Allowable bearing capacity for conventional footings after treatment was 300 kN/m².
> Tests included SPTs and undisturbed samples using a piston sampler for laboratory comparison.

DYNAMIC REPLACEMENT

This is similar to dynamic compaction, but here the tamping is used to form large-diameter granular plugs or stone columns through soft clayey soils and fills to an underlying bearing stratum. The method combines the features of dynamic consolidation with the stone columns formed in the vibro-replacement technique.

Applicable particularly where part of the site is granular rubble fill and part clayey fill to avoid potential differential settlement. The method will cope with obstructions in the fill.

Design, based on relative stiffness, is similar to that for stone columns (6.4). The method tends to provide better improvement at depth than near the surface – say, 50 per cent increase over original stiffness between the columns. High internal shear resistance is built up in the granular columns, which also act as vertical drains to assist consolidation.

The layout of imprints is initially 5–10 m. Typical dimensions are column diameter 3.5 m at 5 m centres, 4 m deep with a 0.75 m compacted raft over the columns.

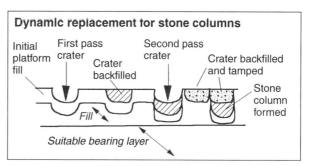

Dynamic replacement for stone columns

Initial platform fill — First pass crater — Crater backfilled — Second pass crater — Crater backfilled and tamped — Stone column formed

Fill

Suitable bearing layer

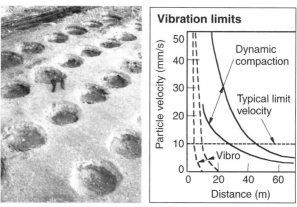

Primary craters

Vibration limits

Particle velocity (mm/s) vs Distance (m): Dynamic compaction, Typical limit velocity, Vibro

Construction methods are similar to dynamic compaction, but a smaller base area tamper is used to facilitate penetration. When the crater is backfilled the tamper pounds this material into the soft ground to deepen the column. Deep intermediate craters are not usually required, but a final carpet tamping is required at the levelled-off surface.

Monitoring is as for dynamic compaction. The effective diameter of the compacted columns and stiffness between columns is determined by pressuremeter tests.

RAPID-IMPACT COMPACTION

The principle is similar to dynamic compaction, but the energy of each drop is less and at a much higher frequency. The hammer hits an anvil plate on the ground surface, not the ground itself.

Suitable for granular soils and fill, like dynamic compaction, but not for natural silts and clays.

Design is based on measuring the improvements in the ground during and after construction – to a specified settlement per blow. Depth of improvement is between 2 and 3 m.

The layout of imprints is typically on a square grid at 2 m centres with secondary tamping between the initial imprints.

Construction uses a specialist heavy crawler rig with a hydraulically operated hammer, capable of around 50 blows per minute from a height of 1 m.

Total energy at each initial imprint is up to 250 tonne-m/m² on a 1.5 m diameter plate. A working platform is placed prior to treatment. Subject to consideration of building vibration and noise, the rig may be operated closer to buildings than cranes for dynamic compaction as less flying debris is produced.

> **Caution**: noise and vibrations are produced during dynamic compaction, and controls and limits are imposed by statute and advised in BS 5228: 1997 Parts 1–4 and BS 6472: 1992.

7 GROUND IMPROVEMENT BY GROUTING

General specifications for grouting conforming to European practice are given in BS EN 12715: 2000 and BS EN 12716: 2001, Rawlings (2000).

7.1 Permeation Grouting

Grouted cut-offs by permeation are considered in 5.1; the following notes amplify the requirements for improving the strength of ground. The process is versatile, but requires skill to achieve optimum results and is highly dependent on the ground investigation providing clear and usable data. The method can be costly when compared with modern jet grouting (7.6) for similar applications.

The principle of permeation grouting is to inject suitable particulate grouts and chemical solutions into porous soil, fill and rock under controlled conditions so that on setting the characteristics of the soil or rock are improved.

As for grout curtains, success in applying this technique depends on many inter-related factors:

- detailed geotechnical, geological and hydrogeological data
- rheological properties of the grouts
- possible environmental constraints
- appreciation of what can be achieved, setting objectives and applying strict controls.

Applications of permeation grouting (in addition to groundwater control) include improving load-bearing capacity of soils, increasing density, underpinning, stabilising open excavations and tunnels, extinguishing tip fires and treatment prior to compensation grouting.

Suitable for treating granular soils as shown in the indicative limits figure.

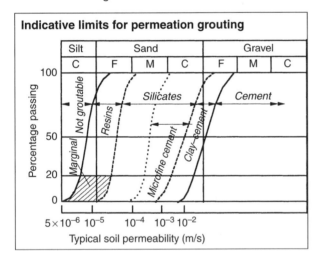

Indicative limits for permeation grouting

Not applicable in silts and clays unless hydro-fracture (7.3) or compensation grouting (7.4) can be used to improve soil strength without causing structural damage. Aggressive or flowing groundwater conditions may cause instability in chemical grouts.

Ground investigation should follow the comprehensive phasing and conceptual ground models recommended. *In situ* permeability and particle/pore size should be thoroughly analysed for the individual strata to be treated. Large-scale pumping tests are not appropriate. Seasonal and tidal fluctuations in groundwater flow and levels and chemical analysis must be determined. The initial strength properties of the ground will be needed in order to determine whether, and to what extent, these properties should or can be improved to satisfy the structural requirements. Post-grouting tests will be compared with these results.

For large-scale applications, field trials of the proposed grouting methods, including zone testing of treated ground, are desirable as part of the investigation.

An environmental impact assessment is necessary to ensure that risk of pollution by grouts is controlled.

A survey of the pre-grouting condition of adjacent buildings is recommended.

Design for improving of the strength characteristics or reducing the permeability of a soil will first consider the relative particle sizes of the soil and the grout – e.g. the groutability index *N*, based on filtration principles, as in J K Mitchell and Katti (1981).

Groutability of soil	Groutability of rock
$N = \dfrac{(D_{15})_{soil}}{(D_{85})_{grout}}$	$N_R = \dfrac{(\text{width of fissure})}{(D_{95})_{grout}}$
Groutable with cement if N is >11 and consistently groutable if >24	Groutable with cement if N_R is >2 and consistently groutable if >5
Groutable with clay–cement if N is > 5	

Permeability of each stratum will be critical to the grout selection as the indicative limits graph and the table show.

Soil	Permeability limits k (m/s)	Grout
Coarse sand and gravel	>1 x 10^{-2} to 5 x 10^{-4}	Cement and clay–cement suspensions
Medium to fine sands	1 x 10^{-3} to 1 x 10^{-5}	Silicate-based gels
Silty sands	>1 x 10^{-6}	Chemical solutions

Depth of treatment required will be related to the stress distribution of the applied load or excavation dimensions.

Grout hole layout (5.1) will depend on:

- overburden pressure at the depth of injection
- porosity of the soil
- grout rheology and setting time
- configuration of the area to be treated.

Spacing of primary holes for permeating sand and gravel between 5 and 25 m deep will be from 1 to 3 m, and <1.5 m in fine sand. Holes are usually on a triangular grid.

The volume of grout is estimated from the volume of pores in a cylinder of soil around the primary injection tube. Injection is stopped when this volume has been placed. Secondary and tertiary volumes and holes for TAM injections will depend on the primary acceptance; some primary grout may be allocated to the secondary holes to reduce travel distance.

Material selection will be based on experience, rheology data and strength requirements.

Cement grout suspensions in water cover the range of cement particle size and chemistry from OP cement to microfine cements depending on the soil being treated, generally producing high *in situ* strength. Additives may be used to reduce the tendency of cement to settle out in 'neat' grouts.

Colloidal suspensions contain very fine particles which tend not to settle out. High-shear colloidal mixers will improve the stability and strength of neat cement grout by thoroughly coating the finer grains with water.

Bentonite (5.2) or other clays are added to improve penetration and reduce bleeding of cement grouts where high strength is not the main consideration; they are also used as fillers in porous soils for groundwater cut-offs.

Bentonite can be strengthened with the addition of sodium silicate to give a permanent gel.

PFA (to BS 3892: 1997) is used as a filler and to improve the suspension of cement and penetration in soil as a result of the rounded shape of the particles. Although PFA is a pozzolanic material, the strength of grout reduces as the amount of PFA included is increased.

Silica fume is also a pozzolanic material with very fine submicron particles which, at low concentrations, helps suspension of cement in high water–cement grouts and improves penetration.

Sodium silicate is used to form soft gels in the soil by reaction with sodium aluminate – mainly for reduction of permeability. Harder gels for ground improvement require higher silicate content with hardening agents, which may increase viscosity and reduce penetration.

Chemical solutions should not contain particles of solids and therefore will penetrate fine soils. The sodium silicate–formamide reaction will produce strong *in situ* grout at reasonably low viscosity; acrylates have low viscosity and will produce medium strength with controllable gel (setting) times.

The toxicity of chemical grouts is low to medium for silicates and acrylates. Acrylamides are highly toxic and now prohibited.

> **Caution**: environmental restrictions and COSSH Regulations apply to all grouting applications. The manufacturers' recommendations for handling chemicals must always be followed.

The strength of grout (UCS) in the soil varies considerably:

- neat cement in sandy gravel   1–10 N/mm²
- silicates and acrylates in sand   0.3–3 N/mm².

Domone (1994) provides data on a range of hardened properties for cement and silicate grouts. Jet grouting (7.6) will produce higher *in situ* strength in most soils.

Viscosity and gel times of many chemical grouts (Karol, 1990) can be designed for injections into fine sands ($k > 10^{-5}$ m/s). Low-viscosity grout (2 cp) with long gel time (60 min) allows greater penetration distance than is possible with short gel time, and at pressures which should avoid rupturing the soil.

Dynamic viscosity v injection pressure

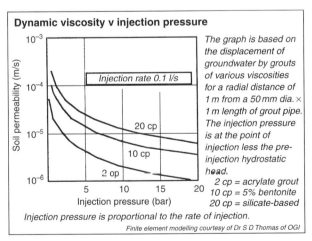

Injection rate 0.1 l/s

20 cp

10 cp

2 cp

Soil permeability (m/s) — 10^{-3}, 10^{-4}, 10^{-5}, 10^{-6}

Injection pressure (bar) — 5, 10, 15, 20

The graph is based on the displacement of groundwater by grouts of various viscosities for a radial distance of 1 m from a 50 mm dia. × 1 m length of grout pipe. The injection pressure is at the point of injection less the pre-injection hydrostatic head.
2 cp = acrylate grout
10 cp = 5% bentonite
20 cp = silicate-based

Injection pressure is proportional to the rate of injection.

Finite element modelling courtesy of Dr S D Thomas of OGI

Methods for ground improvement by permeation are similar to those used for grout curtains (5.1).

Tube à manchette (TAM) system:

- phased injection stages for each stratum
- the specific volume of the appropriate grout formulation in each phase is based on fixed penetration distance from the TAM injection point.

Grout lance injection can produce effective treatment of uniform soils at much lower cost than TAM but without

---

the versatility. The lance may be a steel tube driven or vibrated to depth (with a removable plug) or the drill rods or casings; 5 m depth is a reasonable depth to treat.

Drilling requires temporary casing for placing TAMs (either plastic or steel) and the annular grout. Depths for ground improvement are usually <15 m.

Grout mixing using high-shear colloidal mixers and stirrers is described in 5.1 and 11.2. Accurate batching is important for chemical grouts where gel times are critical. Two-shot processes with short gel times where components are mixed at the hole require continuous metering and proportioning pumps.

Bentonite should be conditioned for several hours before adding cement or chemicals to give a uniform product.

Case: chemical grout underpin in Kirkcaldy

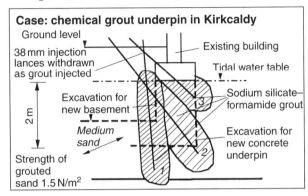

Ground level
38 mm injection lances withdrawn as grout injected
Existing building
Tidal water table
Excavation for new basement
Sodium silicate–formamide grout
2 m
Medium sand
Excavation for new concrete underpin
Strength of grouted sand 1.5 N/m²

Grout injection pressure is proportional to the rate of injection and is also a complex function of yield value and viscosity of the grout and the distance travelled. Simplistically, injection pressure should not exceed overburden pressure, say, 0.2 bar/m depth to avoid rupture and prevent grout taking preferential paths. However, cement grouts are Bingham fluids (5.2) with relatively high yield values and will require higher initial pressure to start grout moving and achieve designed penetration. In open, uniform soil, pressures well in excess of overburden are feasible. Chemical solutions (Newtonian fluids) require lower initial injection pressure (Littlejohn, 1993), but increasing as the grout front gels.

In open soils, high-viscosity grout will be injected initially up to the limiting pressure, followed by thinner solutions as

Pit excavation stabilised with cement–bentonite and chemical grout

needed to infill pores and give desired strength. Injection rates are typically 5–20 l/min and radius of travel is limited to <1.5 m. The sequence of injection should ensure that groundwater can be displaced.

The above assumes that TAM methods are being used.

Pumps (11.3) should provide good control of pressure and flow rates. Long-stroke, double-acting piston pumps are ideal for permeation, but screw pumps are now more common. Output needed will vary from 5 to 50 l/min and from 1 to 20 bar pressure, depending on grout and depth.

Monitoring should record volume, rate of injection and pressure for each lift in TAM grouting. Interpretation of the records will demonstrate whether there is hydro-fracture and whether blockages are occurring due to filtration of grout particles by the soil or increase in viscosity.

Checks on surface movement are essential.

Grout rheology should be checked and modified as needed from the observation of the in-ground performance.

Following treatment, static and dynamic penetrometer testing will indicate strength improvement. Zone tests for bearing capacity are useful.

7.2 Compaction Grouting

The principle of compaction (or displacement) grouting is to improve the bearing capacity and density of the soil by injecting stiff mortar to form a bulb around the injection tube which displaces and compacts weak soil without premature heave of the surface. The mortar should remain as a coherent mass as it is injected and not be capable of entering pores or following preferential paths in the soil, causing hydro-fracture with little compaction.

Hydro-fracture (7.3) or rupture of fine granular soil may, however, be deliberately induced to facilitate intrusion and compaction of soil between lenses of grout and to reduce permeability.

Suitable for treating a wide range of loose granular soils and voided fill. Typical initial SPT *N-values* range from 0 to 15.

Not applicable to saturated silts and clays – unless hydro-fracture techniques can be carefully implemented and pore pressure increases controlled.

Applications include compensating for settlement above tunnels (7.4) and below foundations, mitigating liquefaction and plugging solution features (Francescon and Twine, 1994).

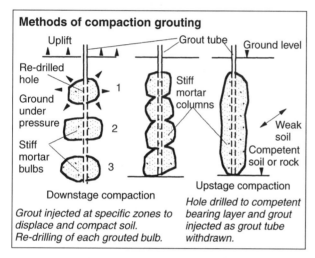

Methods of compaction grouting

Downstage compaction
Grout injected at specific zones to displace and compact soil. Re-drilling of each grouted bulb.

Upstage compaction
Hole drilled to competent bearing layer and grout injected as grout tube withdrawn.

Ground investigation should determine the variations in geotechnical parameters, including density, strength, groundwater conditions and permeability – as for all grouting projects.

A structural survey and risk assessment are essential where surface movement is to be induced or is possible as a result of treatment. Field trials are recommended.

Design is empirical and based on observations and modifications as needed during the treatment. A clear picture of existing conditions and required/likely improvements is essential.

Hydro-fracture or intrusion grouting should be considered in alternating loose and compact strata.

Materials for the stiff mortar are rounded well-graded sand and cement, with 15–20 per cent PFA added and low water:cement ratio; the addition of bentonite will improve pumpability but will increase plasticity. Bleeding should be minimised and the slump should be around 30 mm.

The strength of the mortar is not a major parameter in compacting the soil; 3 N/mm² is typical, higher for structural support as in 'grout jacking' below.

The volume per stage will be set initially as a percentage of the soil volume around the injection tube, say, 5–15 per cent depending on voids, and increased or decreased depending on surface heave observed.

Grout hole layout varies from 1 m centres near the surface to 5 m centres for deeper compaction with over 10 m of confining overburden pressure. Primary and secondary holes are on triangular or rectangular patterns.

Methods used are either:

- *'downstage'*, where the upper layers are treated first, allowing greater pressures at depth with reduced surface heave or
- *'upstage'*, where the grout hole is drilled to full depth initially and the grout injected as the grout pipe is withdrawn in stages.

Drilling by overburden methods using the casing as the injection tube is necessary for the upstage method. For downstage work a casing must be sealed in the top of the hole and each stage has to be re-drilled after injection, making it more expensive. Here, the open hole below the injected zone must be immediately grouted.

Driving or vibrating the tube to depth is also feasible. A reaction against the tendency for the grout pressure to push the grout tube out of the ground may be needed.

Grout mixing is by pan mixers for batch mixing on site or by ready-mix delivery trucks.

Grout injection pressure will be from 5 to 25 bar, depending on surface reaction, but pumping may continue to a set refusal pressure related to the overburden. Filtration of water may occur under excess pressure, resulting in blockages. Caution is needed when injecting large volumes near existing structures and excavations.

Injection rate will depend on the looseness of the soil: in free-draining and dry soil 30–100 l/min at depth; 10–30 l/min in poorly draining soil and near the surface.

If injection pressure falls while pumping at a constant rate, injection should be stopped as compaction is no longer being achieved or grout is probably breaking back to the surface. Once surface uplift starts, little effective compaction of the soil occurs and injection should be stopped. Secondary holes should be used to compact soil between primary holes. Tertiary holes are occasionally used in widely spaced hole layouts.

Pumps are generally concrete pumps with positive feed capable of providing even pressures up to 60 bar but with delivery rates variable from a few litres per minute to 300 l/min maximum.

Grout lines, 50 mm minimum diameter, must be capable of withstanding the high pressures used and resisting wear; high friction losses will occur.

Monitoring of pressure, volume, rate of injection and grout-take should be recorded automatically for each stage. The observational method is appropriate particularly when using electro-levels to control surface movements. CPTs and SPTs of the soil should be taken as treatment progresses and on completion.

Piezometers may be useful in slow-draining soil to decide injection pressure and volumes.

GROUT JACKING

This method also uses stiff mortar mixes for levelling machinery bases, concrete road slabs and differential settlement in buildings, with injection immediately below the structure (King and Bindhoff, 1982). This technique requires a solid bearing zone to provide the reaction to the injection, in some cases requiring preliminary injection at depth.

An investigation into the causes of the differential settlement by structural engineers is essential before treatment, and careful monitoring is required during application. Fluid grout has been used successfully to stabilise concrete road slabs where rocking and 'pumping' were causing damage.

7.3 Hydro-fracture

The principle of hydro-fracture grouting is to open up existing fissures or create fissures by controlled injection of grout under pressure and thereby provide lenses of material to compact and strengthen soil or reduce permeability. Fracturing of the ground (claquage) can occur as a result of poor control of a particular injection process, e.g. formation of lenses of grout when permeation was the aim, ground heave and wastage of grout due to surface break-back.

Suitable for improving strength and stiffness in fine-grained soils, fill and clay. The process is unlikely to be effective in coarse soils, but it can provide some final reduction in permeability in alluvial grout curtains by opening up paths for further permeation.

Applications include reducing permeability, compensating for settlement due to tunnelling, raising and levelling foundations, stabilising slopes, compaction of soil between the fracture zones and improved consistency of clay soil due to ion exchange from the cement grout lenses.

Ground investigations should provide basic geotechnical parameters, particular attention being paid to the state of soil consolidation and the *in situ* stress conditions and permeability of individual strata. Water-pressure tests may assist in determining the appropriate fracture/injection pressure.

Structural surveys and risk assessments are required and field trials are recommended.

Design is based on the use of steel tubes à manchette for repeated injections at each sleeve. The objectives of fracturing the soil must be established and construction parameters set accordingly.

While some progress is being made on the use of finite element methods for compaction and hydro-fracture applications (Nicholson *et al.*, 1994), the method relies mainly on experience and empirical and observational techniques.

Fracture grouting with strong viscous grout to raise structures (Raabe and Esters, 1990) requires structural analysis of the stiffness and loading of the building and a detailed procedure for injection points, depths and pressures, as used for compensation grouting (7.4).

In order to control lateral spread and wastage of grout it may be necessary to install a vertical boundary grout curtain. Similarly, preliminary upper and lower boundaries may need to be injected to provide thrust zones.

Materials used include a range of particulate and chemical grouts with high to low viscosity, depending on applications and objectives.

The degree of fracturing is a function of the grout rheology, injection rate and pressure and stratification, e.g.:

- Thick bentonite–cement grout will tend to provide some ground compaction as the fracturing propagates.
- Chemical grouts with short gel times will be continually fractured as fresh grout is injected through the gelled zone to improve soil strength and reduce permeability (Karol, 1990).
- Hydro-fracture with low-viscosity chemical grout will open fine soil and facilitate permeation.

Grout hole layout will depend on the application and the weight of overburden to resist uplift where improved permeation is the objective, typically 1–3 m for primary holes; secondary holes are dependent on primary injections.

Where raising of a structure is required more uniform uplift may be obtained with deep holes.

Inclined holes and horizontal holes may be used.

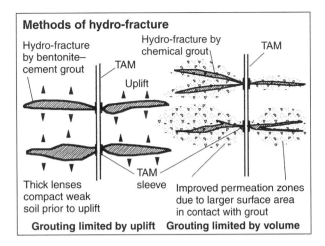

Methods of hydro-fracture

Hydro-fracture by bentonite–cement grout

Hydro-fracture by chemical grout

TAM

Uplift

TAM sleeve

Thick lenses compact weak soil prior to uplift

Improved permeation zones due to larger surface area in contact with grout

Grouting limited by uplift   Grouting limited by volume

Method of treatment is mainly using TAMs so that repeated injections can be made of a variety of grouts. The method is as described under permeation grouting (5.1 and 7.1).

Drilling for TAMs is as described (5.1).

Grout mixing will require colloidal mixers and stirrers, with agitation/holding tanks as needed.

Grout injection pressure and injection rate for a fixed volume of grout through each sleeve will usually be higher for fracturing compared with steady permeation, say, 20 bar after initial fracture at rate of 10 l/min, but care is needed to avoid break-back to the surface.

The number of injections at each sleeve will depend on objectives and observations made during the treatment. Each stage of injection should be allowed to harden before re-injecting that stage.

Grouting is stopped at each stage when:

- the fixed volume of grout has been injected or
- the required lift has been achieved or
- break-back or unwanted surface heave occurs.

Pumps should be high-pressure piston pumps capable of close control up to 50 bar, with compatible pipework. Output again depends on the application.

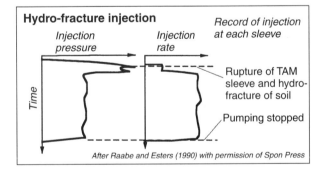

Hydro-fracture injection

Record of injection at each sleeve

Injection pressure

Injection rate

Time

Rupture of TAM sleeve and hydro-fracture of soil

Pumping stopped

After Raabe and Esters (1990) with permission of Spon Press

Monitoring is vital to the success of hydro-fracture. Grout injection pressure, volume and surface movement should be recorded continuously, with immediate intervention by the supervising engineer to make changes in accordance with the set-down procedures. The success of the treatment may be self-evident from the levelling data – the amount of lifting required being achieved or the arresting of settlement.

For other applications, permeability tests and piezometers will be needed; also penetrometer and pressuremeter tests for improved strength.

Coring is not effective as recovery of samples with grout lenses and soil intact is difficult.

7.4 Compensation Grouting

In this *Introduction*, compensation grouting refers to grout injections designed to protect structures from potential damage as a result of adjacent or underground excavation.

The principle of compensation grouting is to inject a sufficient volume of grout to ensure that the building remains at its existing level while excavation takes place below, i.e. to compensate for the change in stresses and ground loss before they influence the structure.

The process will be a combination of techniques:

- controlled hydro-fracture grouting
- compaction grouting; and possibly
- permeation grouting, where feasible, to form a roof against which compaction can thrust or to facilitate hydro-fracture.

It is a flexible process which is controlled by the observational method (1.5) or by prediction and design.

Suitable for fine soils and clay (with hydro-fracture), and coarser soils (with pre-treatment).
Not applicable to fissured rock.

Applications are mainly in urban areas to protect buildings from damage due to settlement during tunnelling; see the discussion of construction of the tunnels for the Jubilee Line Extension in London in Burland *et al.* (2001).

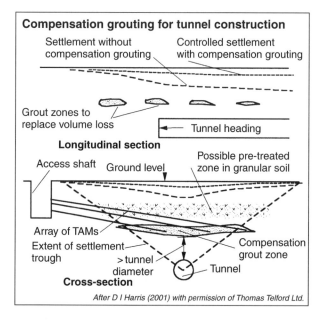

Compensation grouting for tunnel construction

Settlement without compensation grouting

Controlled settlement with compensation grouting

Grout zones to replace volume loss

Tunnel heading

Longitudinal section

Access shaft Ground level

Possible pre-treated zone in granular soil

Array of TAMs

Extent of settlement trough
>tunnel diameter

Compensation grout zone

Tunnel

Cross-section

After D I Harris (2001) with permission of Thomas Telford Ltd.

Ground investigation should provide data in respect of stresses and settlement resulting from underground excavation, particularly on ground stiffness, with additional information relating to the methods of construction and structures to be protected:

- excavation geometry, rate of tunnelling or excavation
- prediction of loss of ground and resulting settlement without treatment
- allowable settlement and distortion with treatment.

The requirements of the risk assessment (1.5) should be considered during the investigation, including the structural survey of buildings above the expected settlement trough.

Design methods now incorporate finite element analysis and modelling as a routine to predict surface settlements due to tunnelling, which are then used as the basis for the initial control parameters for the compensation grouting programme. These predictions have been shown to correlate well with actual measurements of settlement.

The design may allow for 'pre-heave' to be induced in appropriate conditions (i.e. a category of 'very slight' risk of damage) so that when tunnelling arrives at the sensitive area, the settlement due to volume loss will reduce the structure to its original level.

Materials will be as needed for compaction, hydro-fracture or permeation, as given above. Tubes à manchette will usually be steel, with sleeves spaced at 0.5–1 m centres.

Grout mix design has to provide for:

- initial fracturing with a fluid cement–PFA grout; some compaction or intrusion may also be possible depending on viscosity
- compaction with a stiff cement–PFA–bentonite paste – giving high viscosity and low bleed. The slump of typical grout paste is 150 mm.

The volume of grout is estimated from the predicted volume loss at the face, and distributed to individual TAM sleeves for a programme of injections. Actual volume per sleeve will depend on ground replacement required and locations of tunnel crown relative to ground surface and of crown to injection point to avoid unwanted distortions. As with hydro-fracture grouting, it is not possible to predict accurately the spread of compensation grout; hence the volume injected at each phase is limited.

Grout hole layout will depend on the soil, grout, depth of treatment and access for the grout holes – as for all grout injections. The use of computer modelling has improved the design for the multi-array TAM grout pipes for treating ground from shafts and from the surface where only limited access is available (D I Harris, 2001). Grouting parameters are fed into the programme initially to refine the layout and updated regularly as part of the control/monitoring process.

Methods use compaction and fracturing techniques.
Drilling for installing TAMs, inclined or horizontally depending on access available, will require overburden drills in silts and sands and augers in clay. The TAM is inserted and grouted with the annular grout as noted in 5.1. Non-sleeved grout tubes are not suitable for precision compensation. Each TAM should be surveyed and the actual position recorded in the computer programme where possible.

Grout mixing will follow the methods for the particular process as described.

Grout injection pressure and injection rate will be determined by whether fracturing or compaction is the main requirement; the range will be 10–30 bar at 10–20 l/min. A shift plan stating location of injection, mix, volume, pressure, etc. will be drawn up by the engineer and computer controls pre-set on mixers and pumps to ensure that the maximum allowable pressure or injection rate is not exceeded. The record of actual takes and pressures will be examined in conjunction with the surface reactions before deciding on the next phase.

Pumps may be modified concrete pumps as noted for compaction grouting, or smaller, high-pressure, low-volume units which can be computer controlled, able to deliver a range of grout viscosities.

Monitoring is vital to the successful application of compensation grouting in urban areas. Real-time displays of ground and building reaction using inclinometers, piezometers, electro-levels on buildings ($\pm$10 arcsecs) and rod extensometers ($\pm$0.2 mm) in horizontal holes are essential for rapid on-site decision-making. Precise surveying is also undertaken every few hours – using reference points outside the area of influence of the tunnelling.

7.5 Rock Grouting

The principle of rock grouting is to fill fissures and voids with a strong fluid grout which sets to bind blocks and fragments of rock into a more coherent mass with improved strength and stiffness that will resist movement under load and erosion by groundwater.

Suitable for improving most fissured rock, discontinuities and fault zones with fractures >0.5 mm and mass permeability of 5 lugeons, using cement grout.
Also used for reducing groundwater flow into excavations and tunnels and under dams and for infilling voids. Fissures and voids infilled with clay or debris are not treatable unless a substantial amount of the infill can be flushed out and replaced with grout (Ewert, 1996).

Ground investigation must determine, in addition to the geological and hydrogeological conditions at the site, the size, frequency, general structure and orientation of fissures and discontinuities. A high core recovery rate may not be feasible in very broken ground, leading to reliance on remote assessment methods such as water testing, borehole camera inspection, impression packers and geophysics (e.g. seismic velocity). Groundwater chemistry is necessary to ensure that the grout used will not deteriorate. Permeability in rock is usually defined by the lugeon *in situ* water absorption test (1.3), but care is needed in interpreting results (Ewert, 1994).
Site grouting trials are useful as part of the investigation.

Design will focus on the groutability of the fissures (groutability ratio in 7.1) and the potential for improving rock mass properties.
Materials will generally be neat cement grout. Microfine cement may be considered for fissures down to 0.1 mm and some resins may be suitable for strengthening rock in limited conditions, such as cliff stabilisation.

Grout mixes with a low water:cement ratio (say, 1:1) will produce the strongest *in situ* results, but mixes up to 25:1 may be used (Houlsby, 1985). Bleeding of mix water and filtration of particles will affect the penetration of particulate grout by leaving water in the fissure or by blocking flow; long-term durability may be compromised.
Additives, particularly PFA, are used to improve penetration and reduce bleed; bentonite will reduce bleed but also reduce the strength of cement grout. Cement with PFA as a filler (say 1:6 to 1:10) will be used for large voids.
Grout hole layout will depend on area to be treated, depth of treatment and grout rheology. Primary holes at, say, 10 m centres, followed by secondary and tertiary holes, are usually needed to ensure that grout fills the maximum spread of fissures. Care is required to ensure that groundwater (and any bleed water) in the fissures is displaced by the grout.
The area of fissures to be grouted will also depend on the load distribution pattern, e.g. under individual piles, within the 'angle of draw' for old mine workings (8.1) or around the perimeter of tunnels to enable rock bolts to be effective.
Where the rock joints dip steeply it will be necessary to use inclined grout holes to intercept the groutable fissures, but care will be needed while drilling to ensure the drill string does not follow fissures rather than cutting across them.
For grout curtains in rock under dams the width is governed mainly by the hydraulic gradient across the curtain and the state of the rockhead.

Methods of grouting are by stage grouting in open holes or by a form of sleeve tube fixed in the hole:

- **Upstage** grouting requires drilling of an open hole to full depth with a packer placed above each stage in the open hole. It is not suitable for fractured rock.

- **Downstage** grouting progresses with drilling of each 1–3 m stage (depending on the state of the rock) then injecting through a grout connection to the top of the hole or through a packer at the top of each stage.

Upstage	Downstage
Advantages	
Hole drilled to full depth in one move – cheaper	Grouting depth not pre-determined
Avoids requirement to flush out each stage before continuing	Allows stage length to be chosen to avoid hole collapse
Disadvantages	
Grout may break back behind packer, causing jamming in the hole	Requires time for grout to reach initial set in each stage before flushing and drilling next stage
Pressure not held after grout refusal; poorer quality	Repeated drill moves may be required to the hole for each stage
Holes may collapse with zones not treated	

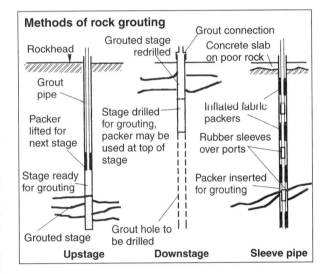

Methods of rock grouting

Upstage / Downstage / Sleeve pipe

Mixing should be with high-speed, high-shear units.
Grout injection pressure required is dependent on the inherent strength of the rock mass and whether there is a need to open fissures to improve the rock strength by grouting. Pressure to start grout flow may be in excess of the limits set by the overburden pressure (say, 0.2 bar/m). This may be acceptable in strong rock mass at depth, but in weak rock at shallow depth, break-back to the surface will occur before useful fissure penetration. Short downstage grouting is indicated and surface caulking of joints or a concrete sealing slab may help limit these losses.
Injection is usually continued until a set refusal pressure is achieved, but the relationship between grout pressure and flow should also be considered.
Pumps are usually of the double-acting piston type with wide pressure range and good pressure control.

Monitoring requires the accurate recording of the grouting parameters with flow meters, pressure transducers and rheology testing. Computer modelling and graphics are now in general use.
Lugeon testing after treatment will indicate ground improvement, and rock coring will provide good visual evidence.

7.6 Jet Grouting

Jet grouting has advanced ground improvement practices significantly in the past 20 years and has displaced injection grouting in many applications – such as reduction of soil permeability, in-ground cut-offs and bearing capacity improvement.

The method is versatile, but it requires expert knowledge to ensure side-effects such as ground heave/settlement or excessive soil overflow do not occur, particularly in urban situations.

The principle of jet grouting is the erosion of the soil by jets of water, compressed air and cement grout to remove some soil and mix the remaining soil *in situ* to form either grouted columns or panels. The jets or 'monitors' (see 11.1) may be self-drilled into a borehole or inserted into a pre-drilled hole depending on the ground conditions and the time and space available. Holes may be vertical or inclined.

Once the monitor reaches the required depth, it is rotated and the jetting fluids are pumped at high pressure to the jetting tip as the monitor is withdrawn from the hole at a controlled rate to form an *in situ* column. Panels or 'wings' can be formed by withdrawing the monitor without rotation. Surplus material flows up the annulus between the tube and sides of the hole and has to be disposed of at the surface.

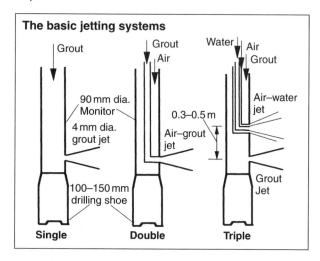

The basic jetting systems

| Grout | Grout / Air | Water / Air / Grout |
| Single | Double | Triple |

90 mm dia. Monitor
4 mm dia. grout jet
100–150 mm drilling shoe
0.3–0.5 m
Air–grout jet
Air–water jet
Grout Jet

Suitable for most granular soils and some clayey soils (shear strength <30 kN/m²). Grading of the soil can be critical – but relative density of the soil is a key factor when the uniformity coefficient is >8.

Typical *N-values* for treatment with the various systems are (see below):

- Cohesive soil   <10 (<5 with the double system)
- Sands and gravel   <50 (double/triple system).

The single system is suitable for sands with $N < 15$.
Large-diameter, strong columns can be constructed in gravels provided grout loss can be minimised in high-permeability strata. Clean sands are easiest to treat, but columns will not be as strong as in gravel.

The column diameter eroded in stiff clays is limited by the energy available, even with the high-pressure triple system. Treatment of peaty soil can be difficult (but see de Paoli *et al.*, 1989) and blockage of the slurry returns is possible in chalk, soft rock and stiff clay.

Not suitable where in-ground obstructions exist.

Applications include foundations for structures, underpinning existing structures, retaining walls and vertical and horizontal in-ground barriers. Encapsulation of landfill contaminants can be achieved provided uncontrolled waste is not brought to the surface.

Ground investigation should determine the detailed geotechnical and geological parameters to produce the ground model and information on adjacent structures. Supplementary investigations will be necessary to assess:

- allowable settlements
- disposal of arisings
- potential size of column and increase in soil strength.

Although soil fabric will be destroyed during jet grouting, it is particularly important to check where major changes in strata occur over the depth to be treated (particularly any cemented layers). The jetting parameters can then be controlled to avoid formation of irregular column diameter and strength variations.

Where variable soil is to be treated, the general disturbance caused by the jetting will overlap thin lenses and there may be no need to alter parameters over the depth of such strata. Field trials are recommended.

Design is essentially empirical based on experience and observation. General specifications conforming to European practice are given in BS EN 12716: 2001.

The guidance in the table is given by A. L. Bell (1993) for the typical improvements in strength and permeability and the diameters possible using the triple-jet system.

Strength and permeability *in situ* depend on the monitor extraction rate, the water jet pressure, the strength of the cement grout used and the degree of soil replacement. In general, the greater the soil volume eroded and replaced by strong cement grout the greater will be the treated strength achieved. In most applications 25 per cent soil replacement is appropriate.

The skin friction of load-bearing columns can be high.

Soil	Compressive strength (N/mm²)	Permeability k (m/s)	Diameter (m)
Sandy gravel	5 to 30	10^{-7} to 10^{-9}	2 to 2.4
Sands	5 to 25	10^{-7} to 10^{-10}	1.4 to 1.8
Silts	4 to 18	10^{-7} to 10^{-10}	0.8 to 1
Clayey silts	0.5 to 8	10^{-7} to 10^{-10}	0.8 to 1

Columns can be reinforced if the steel is pushed/vibrated into fresh material immediately after mixing.

Materials to achieve the figures in the table have water:cement ratios of 0.6–0.8 and a high degree of soil replacement at the higher range of strength. The water:cement ratio is >1.0 at the lower end.

Where reduction in permeability is the main objective a water:cement ratio up to 2 with the addition of up to 10 per cent bentonite and grout strength of around 5 N/mm² will be used. In high-replacement applications, PFA may be used as a filler – but avoid abrasive sand filler.

Grout hole layout depends on the diameter of column achievable so that by using primary, secondary and tertiary columns the columns overlap for areal treatment. The effective diameter of the columns (3 m maximum) depends on:

- injection pressure, flow rate and rheology of the grout
- rotation and withdrawal rate of the monitor
- grading, density and stratification of the soil.

Smaller-diameter columns (<1 m) at close spacing should be used for underpinning applications.

Method selected will depend on the soil conditions:

- **single jet** for mix-in-place application in clean sands, with minimum replacement of soil and least flow-back to the surface; maximum diameter 1 m

- **double jet** produces greater diameter, but the air from the jet gets incorporated into the *in situ* mix, potentially reducing the strength and increasing permeability
- **triple jet** is more complex but more efficient and is now the most widely used monitor, producing the largest diameter of treated ground in the widest range of conditions. The water jet improves the flow of grout. It can be used to erode and replace the *in situ* material with strong grout, but this requires measures to deal with large amounts of spoil.

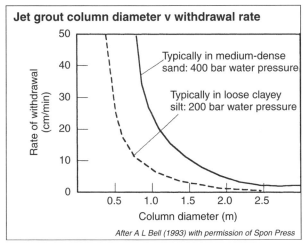

Jet grout column diameter v withdrawal rate

Typically in medium-dense sand: 400 bar water pressure

Typically in loose clayey silt: 200 bar water pressure

After A L Bell (1993) with permission of Spon Press

Drilling of 100–200 mm diameter holes may be with a bit below the monitor, with a pre-drilled cased hole using rotary percussive overburden drills or using reverse-circulation drilling to stabilise an open hole, into which the monitor is placed. Early equipment was based on modified drilling rigs, but purpose-built units which can drill, install the monitor and form the columns (to 25 m deep) with electronic controls and recorders are more efficient. The single system, with only the grout line to connect, will readily take extension rods.

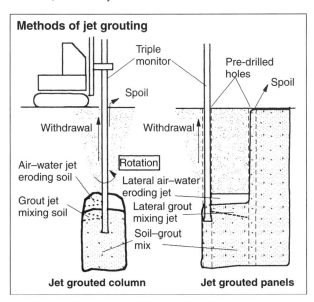

Methods of jet grouting

Triple monitor

Pre-drilled holes

Spoil

Spoil

Withdrawal

Withdrawal

Rotation

Air–water jet eroding soil

Lateral air–water eroding jet

Grout jet mixing soil

Lateral grout mixing jet

Soil–grout mix

Jet grouted column   **Jet grouted panels**

Tolerances for drilling are usually set at 1 in 75, but for in-ground barriers and retaining wall applications it is essential that columns overlap without gaps.
Depth of treatment is governed by the mast height available, as connecting additional lengths of special extension tubes to the triple monitor is not practicable.
Grout mixing requires accurate batching of the grout constituents. Where bentonite is added to the cement grout to reduce permeability of the treated ground the slurry should be conditioned.

The rheology of the mix is important as high-pressure, high-volume pumping is required, which can lead to separation of the mix and blockage.

Typical injection parameters are	
Single system	
Grout injection pressure	300–500 bar
Injection rate	150–500 l/min
Triple system	
Water injection pressure	up to 600 bar
Grout injection pressure	5–30 bar
Air pressure	2–15 bar
Grout injection rate	50–200 l/min

Caution: all jet grouting systems operate at high pressures. The jets should not be operated above the ground except under strictly controlled and supervised conditions.

Injections should be stopped if heave or settlement is observed on the instrumentation (Sheen, 2001–2).
Rotation rate will range from 10 to 15 r.p.m. in granular soils, less in weak cohesive soil.
Withdrawal rate governs the energy applied to the ground and is critical in determining the diameter; it is usually around 250 mm/min for optimum results. Too high a rate may cause 'corkscrewing' of the column.
Surface spoil to be disposed of using the triple system can be >5 m³/h for high-replacement projects. Licences may be required for disposal to special off-site tips.
Pumps may be high-pressure positive displacement screw pumps or triplex pumps for the single system; water pumps are usually triplex piston pumps. For the high air pressures indicated, specialist compressors will be needed.
All delivery lines and valves must be rated for the high pressures.

Monitoring of the process at all stages is critical to success; this includes:

- continuous recording of grout flow rate and pressure, rotation and withdrawal rate (Jameson *et al.*, 1998)
- regular checks on inclination of the monitor tube
- checks on grout rheology and grout cube strength
- checks for signs of heave (particularly in clayey soil)
- condition of arisings from the hole.

Excavation strengthened by jet grout

Tests to verify treatment include:

- core samples of the completed column or panel
- SPT and CPT
- permeability of the column
- zone and column loading tests
- crosshole seismic geophysics.

Where pre-contract trials are required it is desirable to excavate the column as deep as practicable to assess the diameter and strength.
Rotating lasers and electro-levels are useful to monitor heave.

7.7 Other Mix-in-Place Methods

The difference between jet grouting and *in situ* soil mixing is that in soil mixing only minimal replacement of the soil with the grout or slurry takes place. Little spoil is delivered to the surface. The method is largely independent of soil type (boulders and obstructions excepted).

CEMENT STABILISATION

The principle of mixing cement powder or grout directly with soil to improve shear strength and bearing capacity, reduce settlement and encapsulate contaminated waste has been improved in recent years by the technique of 'deep soil mixing' to reach depths of 50 m (Porbaha 1998, 2000; Porbaha *et al.* 1998, 2000).

Suitable for a wide range of soils and fills, including soft clay, alluvium, marine deposits and organic soils. Treatable SPT *N-values* range from 4 in clay and up to 30 in granular soils. Liquid limits between 40 and 140 per cent and water contents in clay up to 150 per cent have been treated, even higher when gypsum is added to the cement.
Organic soils can be stabilised by using specialised cements, but generally slaked lime is used as an additive as for lime piles (below).
Can be used close to buildings – low noise and vibration.
Columns can be vertical or inclined. The ground is not pressurised to any extent; hence heave during construction should not be a problem.
Not suitable where boulders and hard lenses are found.

Applications include settlement control under embankments and light buildings, slope stability, structural pile walls and shaft sinking using interlocking columns, in-ground barriers and cut-offs.
A five-year study of *in situ* mixing to treat contaminated waste is provided by Al-Tabbaa and colleagues (1998a,b, 2000, 2002).

Ground investigation should determine the depth and strength of compressible strata and the presence of boulders. Grading of granular soil will affect the strength of the resulting soil–cement mix. The assessment of soils suitable for stabilisation is given in BS 1924: 1990.
Other parameters required will depend on the prime purpose of the treatment established after the preparation of the ground model – depth of stressed soil, depth to bearing layer, location of slip planes, organic content, contamination levels and designations, groundwater levels and movement.
Detailed permeability of the *in situ* soil for cut-offs using mix-in-place columns is not critical as the soil fabric will be destroyed in forming the barrier.
Site trials are recommended.

Design is based on experience, trials and empirical formula to estimate settlement, bearing capacity and column spacing (Harnan and Iagolnitzer, 1994).
As with the use of jet grouted columns for retaining walls, soil–cement columns are limited by the compressive strength of the column and by whether a reinforcement cage or steel H-beam can be inserted immediately after mixing. The columns must interlock – construction of primary (female) and secondary (male) columns is as for secant pile walls (5.6).
Skin friction, dragdown and the group effect of columns should be assessed for load-bearing columns.

Depth of treatment will be determined by extent of the contamination to be encapsulated, location of the bearing layer or impermeable layer.

Spacing of columns will be on a 3–5 m grid, depending on the applied load and strength of the column. For encapsulation of contamination, fully interlocking columns should cover the area.

Materials for mixing are either cement grout (water:cement ratio 0.4–0.5, depending on the water content of the soil) or dry cement powder for higher *in situ* strength. Gypsum or lime may be added. Whether grout or powder is used, the cement content of the treated ground should be 0.2–0.3 tonne/m^3 of soil.
Typical compressive strength at 28 days will be 1 N/mm^2 in clayey soil to >10 N/mm^2 in clean coarse sand. With dry cement injected pneumatically the soil–cement strength can be doubled for the same cement content.

Construction methods include the following:

Drilling for 600 mm diameter columns is with a plate/mixing auger or with mixing blades at the end of a continuous flight, hollow-stem auger. Crane-mounted units with a kelly bar and turntable can handle plate/mixing augers up to 3 m diameter to 10 m deep.
The deep soil mixing technique for large (3 m) columns and block treatment of landfill waste requires the use of multiple mixing augers and purpose-built high-torque rigs (>150 kNm) to rotate the augers to depths of 40 m. The configuration of the augers and shafts varies, from two to five or six around a common centre.
The auger is screwed into the ground to required depth – without removing any spoil. The auger is then rotated and counter-rotated, raised and lowered for several passes of, say, 1 m, while a predetermined quantity of cement grout is injected through the stem at each pass.
Precautions are needed to avoid bringing contaminants to the surface and releasing airborne particles.

Injection rate for cement grout (mixed in a high-shear mixer) is 50–100 l/min to produce the required strength. The rate must be limited to avoid increasing the *in situ* volume of soil–cement after mixing and compacting with the auger blades. For pneumatically injected cement, delivery rate to the auger, rotation and withdrawal rates are controlled from instrumentation in the rig cab.

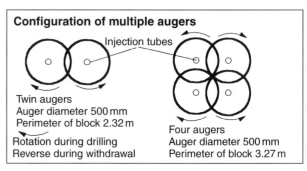

Configuration of multiple augers
Injection tubes

Twin augers
Auger diameter 500 mm
Perimeter of block 2.32 m
Rotation during drilling
Reverse during withdrawal

Four augers
Auger diameter 500 mm
Perimeter of block 3.27 m

Monitoring of the rotation, location in the hole and grout injection by data-loggers and regular testing of the rheology and mixed soil–cement are essential.
Groundwater monitoring wells will be required when encapsulating contaminated landfill and piezometers should be installed to ensure the effectiveness of soil–cement walls as cut-offs.
Pressuremeter tests are useful in and around the columns.
Zone loading tests are suitable for checking bearing and settlement improvement.

LIME COLUMNS

The principle of soil stabilisation using lime, in powdered or slurry form, is the flocculation of the clay content in the soil to increase shear strength and reduce moisture content and swelling potential.
The stabilisation mechanisms include:

- expansion of the column of lime on hydration, causing consolidation of the clay and reduction of the moisture content
- migration of calcium and hydroxyl ions from the lime column, causing flocculation of the clay

Deep soil mixing augers

- increasing strength of the lime column as hydration continues.

Suitable for improving bearing capacity and reducing settlement of soft estuarine clay and silt, provided that the clay content is >20 per cent. The technique is most effective in low-plasticity clay with low initial water content; it is also effective in high-sensitivity clays. For low clay content or organic soils, gypsum, PFA or cement is added with the lime to assist in the ion-exchange reactions. Dry mixing can be effective in peat (Rogers *et al.*, 2001). Lime columns and lime–clay columns will act as drains.

Applications include stabilising embankments, cuttings, shafts and trenches, reducing settlement and increasing the rate of consolidation.

Ground investigation, in addition to the basic data, should determine the clay and organic content, and pH of the soil; the depth, thickness and shear strength of compressible layers; and Atterberg consistency limits. Where contaminated soil and groundwater are to be treated or encapsulated using lime as a neutralising agent, monitoring wells and sampling will be necessary. Onsite trials are usually necessary to assist in estimating the lime content needed and the use of other additives. Laboratory tests should be correlated with field trials.

Design is based on experience and trial mixes for the site. In addition, the structural characteristics of the stabilised soil and the transfer of load between the structure being supported and the composite soil–lime column should be assessed (Broms, 1993, 1994). Individual lime columns can support loads of up to 100 kN, but leaching of lime may occur over time.

Initially there will be some loss of strength in sensitive clay around the column, but after a year the improvement in shear strength of soft clay (<20 kN/m^2) can be considerable (>100 kN/m^2). This will impact skin friction, drag-down and group effects in load-bearing columns.

Depth of treatment (10–40 m, depending on method) should be to a suitable bearing layer, through compressible layers, to support the column load with minimum settlement. The diameter of an augered lime column is typically 200–500 mm. Where greater diameter is required multiple augers are used.

Materials – *unslaked lime* (or 'quicklime') in powdered form is used where the moisture content of the clay is high. The heat of hydration will dry out the clay in time and speed the reaction with the clay.

Slaked lime (hydrated lime) slurry is used for pumping through hollow-stem augers for *in situ* mixing. Typical proportions to dry weight of soil are:

- soft clays: 5 per cent lime, increasing with moisture content and plasticity to 10 per cent
- inorganic clays with low plasticity: 6–8 per cent lime
- organic clays: 3–10 per cent lime with gypsum (3 to 1) added; soil moisture content over 100 per cent
- organic soils (low clay content): lime and PFA.

Construction methods include:

- injecting unslaked lime powder using compressed air through drill rods and mixing blades or paddles as the rods are rotated/withdrawn (the 'Swedish method'); maximum depth is 15 m; diameter is up to 600 mm
- injecting slaked lime slurry through a hollow-stem auger as the auger is rotated/withdrawn; using a large auger head with mixing blades, maximum depth is 30 m; diameter is up to 2 m
- injecting slurry through multiple augers as for soil–cement mixing, or using powdered quicklime injected pneumatically; depth is up to 40 m
- placing unslaked lime into pre-bored holes and compacting *in situ*.

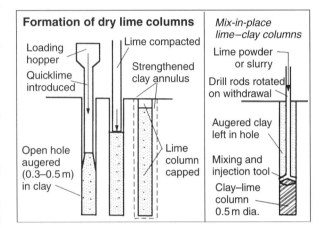

Adjacent columns can be completed consecutively without the need to await gain in strength; this minimises any untreated soil between columns.

An accurate batching plant is required, with slurry mixed in high-shear mixers. No spoil should be brought to the surface and a shroud around the mixing blades will reduce the release of airborne particles.

Lime slurry with the addition of cement and PFA has been used to encapsulate and neutralise contaminated waste *in situ* (Parker *et al.*, 1996). Embankments have been constructed immediately to full height on clay strengthened and drained by lime columns.

Clay slopes are stabilised by lime columns improving the shear strength, the angle of friction and the shear resistance across slip planes.

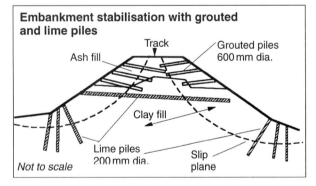

Monitoring will depend on the application:

- lime content tests as in BS 1924: 1990
- triaxial and direct shear tests may be appropriate
- *in situ* loading tests on columns or zone loading tests
- pressuremeter tests in the column and soil
- coring of the column.

Caution: personnel protective equipment, including chemical-resistant splash suits, must be provided when working with quicklime.
Air monitoring may also be required in areas where ventilation is limited.

8 CAVITY STABILISATION

8.1 Old Mine Workings

Old workings exist throughout England, Wales and Scotland, and development of affected areas requires detailed investigation, risk analysis and effective measures to counteract surface subsidence (Healy and Head, 1984). Consolidation of old shallow workings by grout injection into fissures and voids is one of the most widely used geotechnical processes in Britain.

The Stability of old workings depends on the method of mineral or ore extraction.

'Pillar and stall' or 'stoop and room', which replaced the old 'bell pits', is the most common form of partial extraction in old shallow coal workings in Britain. This system consists of sets of roadways driven into a seam at right angles to each other, thereby forming coal pillars to support the roof. Many mines had these pillars removed on abandonment and others were partially stowed with mine waste. Progressive failure of shallow pillar and stall workings due to roof collapse continues for a considerable time after extraction and causes surface disturbance.

'Longwall' mining, mainly employed for deeper mines, extracts more coal by working a straight advancing face, the roof being allowed to collapse or the void stowed with waste to reduce surface movement (Price et al., 1969).

Modern longwall methods produced a 'wave' of surface settlement which kept pace with the advancing coal face, but stabilised when the face passed the site.

A roof fall in shallow workings is the result of:

- collapse of the strata beam spanning between pillars – dependent on lithology, joints and span
- crushing of the pillars – rare at shallow depth
- floor and roof movements.

Span collapse will create a new cavity, which will be stable if a natural beam or arch is formed in strong overlying strata. But cavities and fissures in overstressed weak rock above the collapsed workings may migrate to the surface, causing open 'crown holes', areal depressions, distortion and settlement of buildings and damage to services.

Generally, if the surface is greater than about 6 times seam thickness above the workings, the bulked-up debris from the roof fall will choke the workings, making crown hole formation unlikely.

These mechanisms are not dependent on surface loading, but on the depth, age, size and state of the workings, and the local geology and groundwater conditions. Even lightly loaded low-rise housing will be damaged by subsidence, and measures must be taken to investigate, assess and mitigate the risks to public safety before building in all mining areas.

Other shallow mines encountered in Britain include chalk, gypsum, limestone, oil shale, iron ore and other metalliferous workings, and large-scale salt extraction. Limestone mining of a 3 m thick bed in the West Midlands was also largely by pillar and stall with strong roofs, but severe subsidence has occurred where the workings are as much as 100 m deep.

The most likely cause of collapse of old salt workings produced by 'brining' (pumping in water to dissolve the salt) and by dry mining is the later ingress of groundwater producing classic crown holes.

Phased ground investigation should include the basic conceptual ground model from the known geology and:

- a desk study into whether mining took place, and if so the mining method and location of shafts (as there was no statutory requirement to deposit records prior to the 1870s, archive searches may be fruitless)
- detailed reconnaissance of the site and surrounding area for signs of crown holes, shafts, outcrops, damage to property or influence of groundwater
- probe drilling supplemented with rotary core drilling to determine the state of the workings and roof
- CCTV examination of open workings
- geophysical surveys, which are limited to detecting voids at shallow depth; crosshole seismic tomography is useful but costly; microgravity and magnetometer surveys also have special applications
- a risk assessment of the effects of the workings on the stability of new buildings by an experienced mining engineer (K W Cole, 1993; K W Cole and Statham, 1992a and b)
- an estimate of the void volumes of wide roadways, collapsed zones and back-stowage in the workings.

Depth of investigation in mining areas should be a minimum of 20 m and down to 100 m in critical areas. The number of holes depends on the complexity of the new works – say, four cored holes (76 mm) for a high-rise block to be founded on rock over workings.

As the early methods of extraction were often random, investigation by drilling methods into the extent of the workings to assess treatment is not always reliable. The geophysical methods noted may produce more useful results. Later pillar and stall workings can be defined reasonably well by probe drilling.

The principle of stabilisation is to prevent further collapse of the workings and resulting movement of buildings by structural means and/or grouting methods.

Large-diameter piles or piers drilled and cased through shallow workings are feasible, with the casing left in place for concrete piles cast *in situ* or removed after inserting pre-cast piles. Lateral movement and dragdown on the piles should be assessed. Piles should not terminate above the roof of old workings unless an adequate thickness of intact sound rock forms a stable roof. Where workings are near to the surface, as in the old bell pits and adit mines, the simplest way of ensuring stability is to excavate to a level below the pit and backfill or use the space for basements founded below the workings. Stiff raft

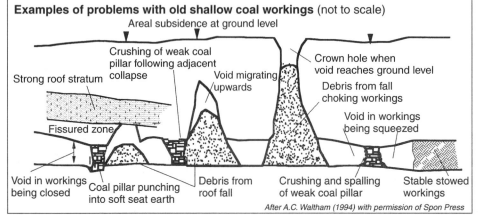

Examples of problems with old shallow coal workings (not to scale)

Areal subsidence at ground level

Crushing of weak coal pillar following adjacent collapse

Strong roof stratum

Void migrating upwards

Crown hole when void reaches ground level

Debris from fall choking workings

Void in workings being squeezed

Fissured zone

Void in workings being closed

Coal pillar punching into soft seat earth

Debris from roof fall

Crushing and spalling of weak coal pillar

Stable stowed workings

After A.C. Waltham (1994) with permission of Spon Press

foundations may be constructed at the surface to bridge areas liable to collapse.

The most common form of treatment is grouting of the voids in open and collapsed workings to prevent surface movement and allow conventional shallow foundations to be used with safety.

It is also possible to alleviate mining settlement to existing structures by grout jacking (7.2).

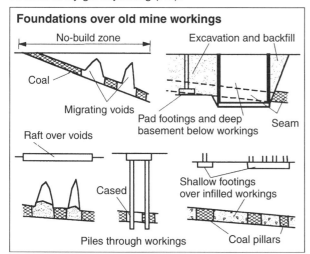

Foundations over old mine workings

Design for infilling by grout is mainly based on empirical rules:

- infill all voids found down to a depth of 20 m below ground level
- infill all voids within a depth of 10 times the seam thickness or
- no treatment if no surface movement has occurred in the past 100 years.

It is rarely possible to undertake investigations from the surface at reasonable cost which will determine the location and volume of all the cavities. The foundation should therefore be of a type which can be modified if further information on cavities is discovered during construction.

When cavities have been infilled it is usual to keep the spread footings as near to the surface as possible to maximise dispersion of the load at workings level.

The area to be treated by infilling is usually determined by the chosen depth of treatment and an assumed angle of draw around the building – typically 20–30°. This area may be excessive if the strains in the strata above partially extracted workings can be shown, as part of the risk assessment, to affect an area smaller than the

underlying voids (Higginbottom, 1987). However, these 'safe depth' and 'safe area' rules have been satisfactory in most cases.

The volume of voids for estimating grout quantities is based on allowing for 60–70 per cent extraction of each seam with 50–60 per cent of that extraction being backfilled (stowed or debris), i.e. approximate grout volume is 25–35 per cent of total seam volume.

There is evidence from grout-take data (Statham and Scott, 1994) that groutable voids are generally less than estimated from these approximations due to dense stowage during mining and debris filling collapsed workings, indicating that grouting in such cases may have contributed little to the long-term stability.

Construction methods

Grout hole drilling involves:

- 100 mm diameter grout holes at 1.5 m centres to form a perimeter barrier around the area of each seam
- 50 mm diameter infill grout holes in a primary grid at 6 m centres followed by a secondary grid at 3 m and tertiary or test holes where appropriate, all with 25 mm diameter medium-density polyethylene (MDPE) grout pipe inserted
- casing through overburden and any upper seams in order to install the grout pipes
- drilling separate grout holes into each seam if large voids are found.

Grouting procedure:

- **A barrier** is formed around the perimeter either by placing pea gravel (10 mm single size) down the 100 mm hole where large cavities are likely or by injecting 3:1 sand:cement grout with some PFA into voids through 50 mm diameter grout pipe. The barrier may not be needed at the up-dip end of the seam. If the workings are flooded, water must be allowed to drain from the treated area for optimum penetration of grout.
- **Infill grout** is usually a mix of PFA and OP cement ranging from 6:1 to 20:1 by weight, with a water:solids ratio of 0.6–0.8 to achieve some limited penetration into debris. The minimum strength specified ranges from 1 to 4 N/mm^2 at 28 days; bleed capability should be <5 per cent by volume.
- **Mixing** of sand–cement grout will require double-drum colloidal mixers, but infill grout is mainly mixed in large re-circulating jet mixers based on centrifugal pumps built to handle solids. Such units can continuously mix grout and are controlled by microprocessors – 400 tonnes of grout per day is typical.

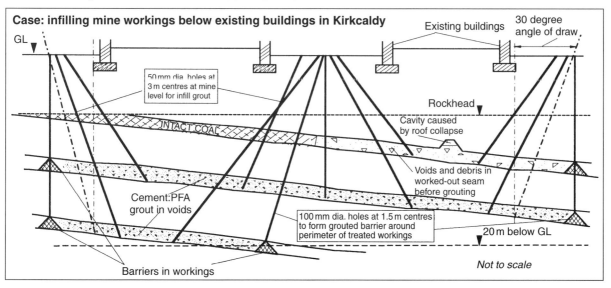

Case: infilling mine workings below existing buildings in Kirkcaldy

Not to scale

- **Grout-takes** will be selective, e.g. 65 per cent of infill holes only accepting 1 m³ of grout, 25 per cent accepting <10 m³ and the remaining 10 per cent accepting >10 m³, being 95 per cent of total volume (Statham and Scott, 1994).
- **Injection** should commence at the lowest (down-dip) level of the seam. Packers are not usually required. Grout pipes are moved up the hole once a set limiting pressure is achieved (say, 0.2 bar/m) to a level where flow will occur, and injection is usually continued until grout overflows the hole. Limiting flow in the cavities by limiting the volume injected at each hole per day seems unnecessary except in steeply dipping strata.

The grout mix may be revised if takes are high (grout flowing away from the treated area) by adding up to 50 per cent rounded sand to the PFA:cement base or adding accelerators to limit flow.

- **Injection pressure** will be nominal for open workings. Pressure grouting with packers in tertiary holes to top up voids or inject fissures above the workings should be carried out as a final check on infilling. If tertiary holes are not required, final holes should be pressure tested with packers.

Caution: methane ('fire damp') may be present and at concentrations of 5–15 per cent by volume in air is explosive. Check for methane every 1–5 m depending on the potential hazard assessment. Drilling should be stopped if a concentration of 3 per cent is encountered in the open. Only restart when ventilation or purging reduces the concentration (ICE, 1993b; BDA, 2002). Smoking must be prohibited.

Schematic layout for high-volume mine infill

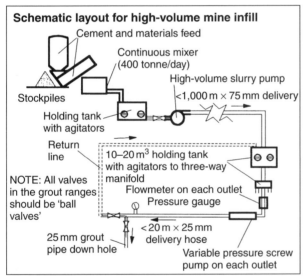

- **Pumping distances** may be up to 1,500 m from mixer to injection point for PFA:cement grouts using high-pressure rotary screw pumps or reciprocating pumps – subject to acceptable rheological properties (viscosity, setting time, particle size) and pipe sizes. Sand–cement grouts are difficult to pump further than 100 m due to separation of constituents and line blockages (Littlejohn and Waterhouse, 1990).

Grout materials:
- The usual materials specifications for cement, sand and PFA apply.
- 'Fly ash' may refer to ashes which have not been produced by the burning of hard coal and therefore have different properties from PFA.
- PFA from storage lagoons will usually be coarser and have higher moisture content than the 'conditioned hopper PFA' generally specified for grouting.

- 'Lagoon ash' and coal waste have been successfully used with low (2 per cent) cement content to produce a cheap, pumpable 'paste' to infill large limestone workings (Jarvis and Brooks, 1996).

Monitoring and verification

By adopting primary and secondary grouting, the infilling progress can be checked during the works. On completion

- separate test holes should be drilled and pressure tested with either water or grout
- coring of the grouted ground and workings must be delayed until the grout has reached adequate strength; it is not feasible for low-cement-content pastes
- CCTV inspection of the treated workings gives good results.

It must be accepted that grouting methods will not completely fill all voids produced by mining activity; infilling must be sufficient to prevent future migration of any remaining voids.

Other stabilising techniques

Where workings are safely accessible and dry, an underground survey can be made. Barriers can be built in the workings as required and low-strength PFA:cement paste pumped in behind the barriers and selectively located bulkheads to control flow.

Alternatively, these workings can be stowed using the modern pneumatic methods employed in active mines.

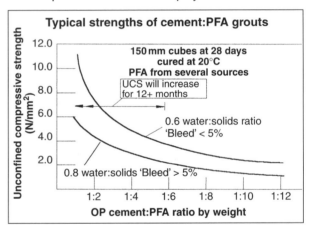

Typical strengths of cement:PFA grouts

Non-cementitious materials such as pea gravel have been flushed down perimeter holes to form barriers across roadways in workings. It may be necessary to spread the cones of gravel formed in order to avoid gaps between adjacent cones near the roof. Bulk infilling by flushing coarse sand down boreholes into the workings has also been successful.

Case: stabilisation of old stone workings under Bath
The roof of these extensive (400,000 m³) eighteenth-century workings can be as little as 2 m below the city streets and has deteriorated considerably over the years, particularly as a result of water ingress. Bulk infilling would cause significant disruption with materials deliveries, hence a programme of new roof support by propping using wooden posts, beams and wedges with steel under surface roads, installed by experienced miners to produce a safe access route, is under way (2002) as a first stage. In hazardous areas infill has been carried out using foamed concrete, and grout bags to replace pillars.

Caution: prospective developers in coal mining areas should consult the local authority and the Coal Authority before undertaking exploratory, stabilisation or building works.

8.2 Mine Shafts

Stability

The hazards presented by old mine shafts cannot be overstated and a comprehensive risk analysis must be undertaken before any exploratory work to locate shafts is started. Risks include:

- collapse of infilling following deterioration of a supporting structure above the base or erosion of infill by mine water
- subsidence of poorly compacted fill
- collapse of the shaft lining,bringing in overburden
- fluidised infill suddenly dropping down the shaft
- rising groundwater driving mine gases up the shaft.

The numbers, sizes and locations of shafts used to access and ventilate old workings are highly variable and collapse of even a small shaft can cause considerable subsidence and damage. If the age, pattern and depth of workings are known, an experienced mining engineer may be able to infer the type and size of shafts in an area.

Diameters of abandoned shafts which may require stabilisation vary from 1 to 7 m, and depths from 20 to 800 m, the shallower shafts being generally the oldest. Early shafts are likely to be clustered, with several shafts in close proximity (e.g. bell pits within 3 m).

Local authorities have strict procedures controlling erection of buildings in the immediate vicinity of abandoned shafts and adits.

The principle of stabilisation is to prevent collapse of the lining and infill by structural support and/or consolidation of the infill and voids.

Ground investigation should commence with the desk study and site reconnaissance as for mine workings.

Field investigation using trial pits and trenches is most effective to determine the location and condition of shafts where overburden is less than 5 m deep. Elsewhere, use closely spaced probe drilling with vertical and inclined holes on a grid starting at the indicated centre of the shaft, supplemented by ground probing radar.

The state of the upper part of the shaft lining is particularly important as collapse may allow loose overburden to flow into an empty shaft.

Levels of groundwater in and near the shaft must be located as water entering a filled or partially filled shaft will result in collapse or settlement of the fill and possibly failure of the lining.

The initial risk assessment will be amended to address the actual shaft conditions found.

Caution: when drilling for investigation or grouting of shafts extreme care is necessary: use widespread grillage and a strong, anchored geogrid over the area of influence of the shaft to prevent drillers and the rig falling into holes caused by collapsing fill or failed linings. Drillers should wear a harness attached to a safety line (BDA, 2002).

Ensure that the flushing medium is selected to limit disturbance while drilling.

Obstructions to drilling such as large timber baulks and old mine cars dumped in the shaft are a hazard – usually requiring re-drilling.

Methane may have accumulated and measures as for mine infilling (8.1) should be adopted.

Methods of treatment may be by excavation, but are usually by infilling with lightweight concrete, gravel or pea gravel, or by grouting of the *in situ* or new fill and structural capping (Healy and Head, 1984). Typical measures are as follows.

Grouting and capping is the preferred treatment in urban areas where buildings may be constructed in the vicinity of shafts. The grouting crew should be on site when the probe drilling is being carried out so that when the shaft is located injection can follow on immediately to avoid potential collapse. Estimating voids from the investigation is difficult; indicators are obvious drops of the drill string or loss of flushing medium.

One central, heavy-duty steel grout pipe is used for shafts up to 2.5 m diameter – four grout pipes for a 5 m diameter shaft. The holes should be taken to the base of the shaft or at least to 250 m. This requires high pull-out capability in the drill rig and strong support grillage over the shaft and surrounding area.

PFA:cement grout (10:1 mix or stronger) with good fluid properties is injected in 2–3 m lifts as the grout pipe is withdrawn. It may be necessary to use TAM methods (5.1) to improve and control grout-takes. Compaction grouting with grout paste is suitable in low-void fills. Grout-takes and depths should be recorded.

Infilling of open shafts where rockhead is near the surface using large cobbles and boulders at the base of the shaft to allow mine water to flow, followed by topping up with clean gravel, is effective. Where 'insets' (intermediate openings) occur in the shaft the open cobble and boulder fill must straddle the opening.

Removing unstable backfill by grab and replacing with granular material may be feasible if the shaft is partially filled over a sound platform. Drilling below the platform (if possible) and filling voids with grout or pea gravel will improve stability.

A reinforced concrete (RC) capping slab with top and bottom reinforcement can be used:

- at sound rockhead where excavation is feasible
- at ground level if overburden is deep, and a heavy-duty geogrid is secured at ground level covering the potential zone of influence.

The shaft should be grouted as above.

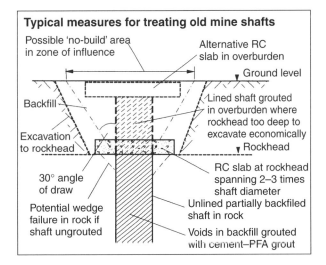

Typical measures for treating old mine shafts

Possible 'no-build' area in zone of influence

Alternative RC slab in overburden

Ground level

Backfill

Lined shaft grouted in overburden where rockhead too deep to excavate economically

Excavation to rockhead

Rockhead

30° angle of draw

RC slab at rockhead spanning 2–3 times shaft diameter

Potential wedge failure in rock if shaft ungrouted

Unlined partially backfiled shaft in rock

Voids in backfill grouted with cement–PFA grout

A bored pile wall around the shaft at a safe distance (2–3 times shaft diameter) bearing on sound rock can be used in deep, weak overburden or where the lining is likely to collapse. An RC slab spanning the wall caps the shaft. The piles should not stress the roof of any shallow workings.

A 'no-build zone' around the shaft equal to the depth of overburden – up to 30 m from the shaft for high-rise buildings (Price *et al*., 1969) – is necessary where treatment may not be fully effective in securing the shaft.

Monitoring and verification

May be by core drilling once the grout has hardened, but more conveniently by test holes and pressure grouting.

8.3 Solution Cavities

Stability

Dissolution of carbonate rocks such as limestone and chalk by rainwater and acidic groundwater opens up fissures to form underground water courses and continuous cave systems which in most limestones are stable within inherently strong rock. While deep stable limestone cavities usually have little effect on surface structures, the associated shallow features of highly variable rockhead, transported debris infilling, gorges and swallow holes give rise to typical karst terrain.

KARST

Karstic features (limited in Britain) are produced by solution of limestone along joints at rockhead, hidden below transported overburden, and within the upper layers of rock, causing considerable difficulties for the engineer:

- large caverns extending several hundred metres, partially or completely filled with sand, gravel and clay, possibly interconnected and containing flowing water
- unconnected smaller cavities, also filled or open
- pipe-like channels within sound rock
- open or filled joints of various width
- pinnacles of limestone beneath overburden
- solution sinkholes (also known as swallow holes, dene holes); collapsed, choked and buried sinkholes.

Most authorities emphasise the importance of dissolution in forming karst features (White, 1990), but karstic landscapes also form in a wide range of rock types which would normally not be considered sufficiently soluble – e.g. where subsurface drainage systems cause erosion and flow through caves or pipes to emerge as springs.

Karstic areas therefore require that particular attention be paid to the secondary geology to highlight requirements for investigation, classification, design and construction (Waltham and Fookes, 2003) and the risk assessment.

Ground investigation of natural voids is more difficult than for mine workings as surface features follow no obvious pattern to aid deduction. Wider geological considerations are therefore important, based on the desk study, vertical and inclined rotary coring and probes on a closely spaced grid to produce the conceptual model. Water pressure tests and records of the dropping of the drill string and loss of flushing water are useful indicators.

Methods of treatment include construction of RC slabs to span the cavities or pinnacles, adits or shafts to access open dry caverns and then infilling with mass concrete. Removal of loose debris from large water-filled caverns is difficult, but where wash-out can be achieved through boreholes, the cavern is filled with concrete or gravel then grouted.

Grouting from the surface, as for mine stabilisation, or from access galleries is difficult but is frequently the only means of treatment; success will vary considerably, depending on the degree of karsticity.

For groundwater exclusion, grouting may have to be preceded by flushing out the debris and overburden clogging small caves and fissures – a time-consuming and costly process requiring high-pressure (>10 bar) air–water jets delivering 150 l/min in holes at 1 m centres (less for 'young' karst) (Kutzner, 1996). The flushed products have to be disposed of under environmental restrictions (i.e. usually not to surface water courses).

Infilling of smaller cavities using PFA–cement grout (with an accelerator to reduce flow) can then be attempted, followed by fissure grouting with neat cement through the cleaned flush holes. This will be a 'hit-and-miss' process; the removal of all the debris from cavities and clay from fissures and consequent bonding of the blocky structure by grout should not be expected.

Pre-treatment for piling in near-surface karstic areas is usually necessary using probing, flushing and fissure grouting below each pile. Where debris-filled voids are encountered, flushing and grouting may be attempted, but it is unlikely that adequate grout will be injected to allow a stable open hole to be drilled for the pile. Simultaneous drilling and casing of the piles will be necessary. Sufficient grout infilling may be possible to avoid loss of pile concrete as the temporary casing is withdrawn. Piles should be socketed into sound rock to achieve bearing capacity.

Jet grouting (7.6) below each pile toe position or shallow foundation can effectively mix-in-place or replace the debris in cavities with grout in the bearing zone. It is important to have an adequate pathway for the venting of debris and to match withdrawal rates to water pressure and flow (up to 450 bar and 80 l/min) and grout pressure and flow (450 bar and 350 l/min) (Bruce et al., 2001a). It is advisable to pre-drill the grout holes into the debris-filled voids in order to insert the monitor.

Compaction grouting (7.2) of loose infill below shallow foundations is also feasible, preferably down to sound rock.

Curtain grouting for a cut-off in karstic features can be difficult to achieve and costly. Bruce et al. (2001b) describe a series of measures including compaction grouting, geotextile grout bags, a multiple packer sleeve system also using grout bags, hot bitumen in high-water-flow zones

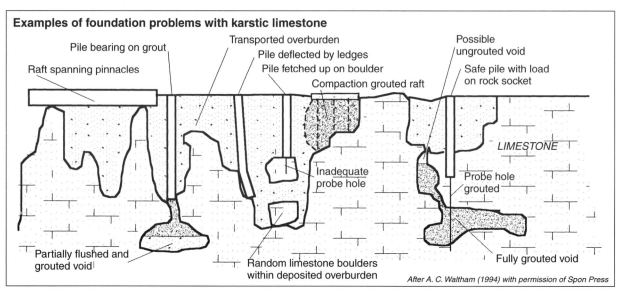

Examples of foundation problems with karstic limestone

Raft spanning pinnacles

Pile bearing on grout

Transported overburden

Pile deflected by ledges

Pile fetched up on boulder

Compaction grouted raft

Possible ungrouted void

Safe pile with load on rock socket

LIMESTONE

Inadequate probe hole

Probe hole grouted

Partially flushed and grouted void

Random limestone boulders within deposited overburden

Fully grouted void

After A. C. Waltham (1994) with permission of Spon Press

and accelerated cement grout to seal water flow in karst to a depth of 67 m.

For dam abutments in karst it has been shown (Ewert, 1996) that following mix-in-place techniques to stabilise infill, grouting of partially infilled fissures can effectively reduce permeability – provided high hydro-fracture pressures (>30 bar) and close centres for grout injection can be used. Trial grouting as part of the investigation is advisable.

Other foundation measures include:

- pile-supported rafts
- drilled shafts, caissons or piers through large sink holes
- mini-piles
- ground beams, bridging rafts
- geosynthetics-reinforced fill beneath footings.

CHALK

Chalk is a porous, soft white limestone made of fine grains of calcium carbonate with joints and bedding planes which can develop as dissolution features – but does not exhibit typical karst topography. Underground caverns are rare as the roof will be weak, although a variety of dissolution hollows, sinkholes, pipes and old shallow mines (for flints and lime-burning) are common.

Case: treatment of sinkhole in volcanic soil near Torquay

Following heavy rain a large sinkhole opened in an estate road; the settlement damaged a water main, which caused further subsidence. A 3 m grid of cased holes was drilled to rockhead, or to 20 m where rockhead was not reached at the centre of the sink, and high-pressure compaction grouting was carried out with a stiff (100 mm slump) sand–cement grout. Following penetration tests, additional injections were made until consolidation and excess water in the coarse *in situ* material was drained.

Typical sinkhole developments

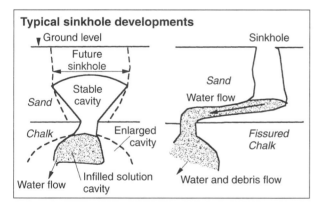

The sinkholes are formed at the surface when the loose infill in the hollows and pipes from past drainage is washed further into the fissures and voids, as these are deepened with ongoing surface drainage. Collapse will also occur due to erosion of weak infill, producing much wider depressions than crown holes or sinkholes from mining (McDowell and Poulson, 1996).

The mines for flints may be only 2–3 m below ground level.

Ground investigation should determine the stratification and lithology of the chalk and the degree of weathering (1.4) to indicate possible dissolution features. Where shallow mining is likely, the use of ground-probing radar can delineate cavities, but a series of vertical and inclined probing is usually required.

As for karst, the risk assessment should include the likely effects of the secondary geology.

Methods of treatment include:

- excavation and backfilling with graded filter material, for near-surface solution cavities
- infill and fissure grouting, as for limestone cavities, but water flushing should be kept to a minimum
- grouting of backfill rubble
- compaction grouting in overburden above a pipe, which can effectively plug the entrance to the enlarged fissure if its location can be determined.

Heavy structures may need piling or caissons taken below the solution features.

SALT

Salt lies at shallow (30 m) depth in the Cheshire salt field, covered by weak overburden. High extraction was achieved historically by mining, and on abandonment, water entered, saturating then dissolving the brine pillars, which collapsed into sinkholes.

Brining operations include pumping the stable, saturated brine lying in the lower part of thick impermeable salt seams and using injected fresh water to dissolve the dry salt. Both can create large uncontrolled caverns which collapse prior to the brine becoming saturated.

Abandoned wells can allow fresh groundwater to enter the cavities, extending the dissolution and potential for collapse.

Case: treatment of old chalk workings near Reading

The deterioration of old chalk workings has resulted in roof collapse and the formation of sinkholes at ground level over many years. The cavities have generally been treated by a combination of infill and injection grouting. However, the local authority decided that some larger cavities, which would be uneconomical to infill but could be accessed, should be preserved. These were stabilised by applying reinforced shotcrete to the sides and roofs of the workings to form structural arches.

OTHER CAUSES OF SUBSIDENCE

Sinkhole in Chalk

There are numerous examples of the collapse of weak surface deposits due to underground erosion as a result of leaking water pipes and sewers, strata separation due to excavations, water entering tension cracks, vibrations from traffic and construction work, inundation by flooding, etc.

Site investigation should determine the root cause of the subsidence and the local geology.

Methods of treatment include the range of construction and geotechnical measures noted above, concentrating on addressing the cause of the problem. Excavation to a sound formation and backfilling are usual for shallow holes.

9 GROUND ANCHORS

Ground anchors are used for a variety of applications in mining and civil engineering to stabilise rock faces and resist uplift and overturning forces acting on structures. They range from short rock bolts for tunnels to long, high-load, multi-strand anchors for dams.

The anchors dealt with in this *Introduction* derive their load-bearing characteristics mainly from the bond or frictional resistance developed between the ground and the grout surrounding the anchor 'tendon' (the steel bar or strand which is put into tension to transfer structural load to the ground) and between the grout and the tendon.

The capacity of the anchor therefore depends on:

- the shear strength of the soil or rock into which the anchor is fixed
- the tensile strength of the tendon and
- the tensile/bond strength of the grout.

The bond between grout and tendon is the main parameter in strong rock, whereas in weak rock and soils the bond between grout and ground will control effectiveness. The durability of the anchor also depends on the groundwater conditions and the corrosion protection provided around the tendon and at the anchor head.

9.1 Tie-back Anchors for Retaining Walls

The principle of using tie-back anchors to support earth-retaining walls is essentially to remove the need for struts or shores within the area to be excavated.

The anchors are tensioned and the load locked in order to limit lateral movement of the wall.

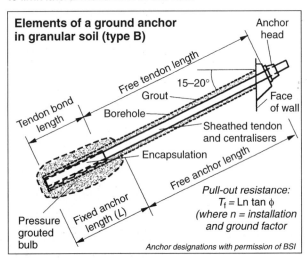

Elements of a ground anchor in granular soil (type B)

Anchor head — Free tendon length — 15–20° — Grout — Face of wall — Borehole — Sheathed tendon and centralisers — Tendon bond length — Encapsulation — Free anchor length — Fixed anchor length (L) — Pressure grouted bulb

Pull-out resistance:
$T_f = Ln \tan \phi$
(where n = installation and ground factor

Anchor designations with permission of BSI

Suitable for resisting lateral movement of exposed embedded retaining walls where adequate resistance to slipping or yielding of the soil or rock mass in the anchor zone is available.

The anchor system may be required for temporary or permanent support for the structure, or as remedial works. The highest anchor loads are achieved in sound rock; but dense granular soil can provide ultimate capacity up to 2,000 kN. Anchor capacity in stiff clays depends on the soil shear strength and whether fracture grouting or under-reaming is carried out to improve the pull-out resistance. Note that permission must be obtained to drill and grout inclined anchors under adjacent property.

Ground investigation should be comprehensive and include desk studies, phased investigations and conceptual models for most applications.

There are two aspects to be studied:

- detailed geological structure (vertical and lateral), fabric, lithology and geotechnical strength parameters of the mass of soil or rock which will carry the anchor load
- parameters for the stability of the wall – anchor system, settlement, groundwater (including seasonal variations), permeability and chemical attack on the anchor components.

Where rock anchors are indicated, then stratification, fissuring and rock quality designation (1.4) have to be determined as well as the rock mass strength. *In situ* testing using pressuremeters and anchor pull-out tests on critical sites are used for design.

Design requirements using mainly empirical methods are covered in BS 8081: 1989. The length of anchor and fixed anchor length to provide the required load are calculated using the recommended minimum factors of safety: 2.0 for temporary anchors (say, 12 months) and 2.5–3.0 for permanent anchors, depending on load and consequences of failure.

The design of the wall and the end bearing must take account of the vertical component of the anchor load. Depending on the type of wall, continuous walings will be needed to distribute the anchor loads – e.g. external horizontal beams for sheet pile walls, internal reinforcement for concrete diaphragm walls

Anchor types as defined in BS 8081: 1989 will dictate the method of design, installation and pull-out resistance.

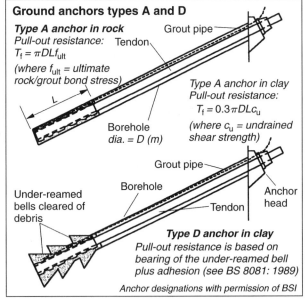

Ground anchors types A and D

Type A anchor in rock
Pull-out resistance:
$T_f = \pi DL f_{ult}$
(where f_{ult} = ultimate rock/grout bond stress)

Grout pipe — Tendon — L — Borehole dia. = D (m)

Type A anchor in clay
Pull-out resistance:
$T_f = 0.3 \pi DL c_u$
(where c_u = undrained shear strength)

Grout pipe — Borehole — Anchor head — Tendon — Under-reamed bells cleared of debris

Type D anchor in clay
Pull-out resistance is based on bearing of the under-reamed bell plus adhesion (see BS 8081: 1989)

Anchor designations with permission of BSI

- **Type A** are typical rock anchors with the fixed end grouted with cement or proprietary resin capsules. They may also be used in stiff clay. Pull-out resistance T_f depends on the bond between the sides of the straight-sided borehole and the grout. Ultimate bond strength is usually taken as 10 per cent of the UCS of the grout (at 28 days), to a maximum of 4 N/mm².

- **Type B** are pressure-grouted anchors used in granular soils and weak rock. The grout permeates or compacts the soil surrounding the anchor zone, producing an enlarged end bulb which provides both side shear and end resistance to pull-out. Pull-out resistance also takes account of drilling method, grout pressure (by applying a factor n) and the effective angle of friction of the soil. n varies from 130 kN/m in fine sands to 600 kN/m in coarse sands and gravel, and can be increased where an enlarged end bulb is formed.

- **Type C** are anchors grouted at high pressure to cause hydro-fracture of the surrounding fine-grained soil; they are typically installed using tube à manchette techniques (5.1). Design has to be based on pull-out tests on site.
- **Type D** are used mainly in cohesive soils where the stability of the under-reamed bells can be assured; say, clays with $c_u > 80$ kN/m^2.

Fixed anchor lengths of between 3 and 10 m are recommended in granular and cohesive soils; >2 m in rock.

Tendon bond length is based on a working bond stress of 1 N/mm^2 for plain steel bars to 3 N/mm^2 for deformed bars where a degree of mechanical locking is possible, using cement grout with UCS > 30 N/mm^2. The length of resin-bonded anchors is based on manufacturers' data or site testing.

Free anchor length must ensure that the fixed anchor is placed beyond the failure plane as determined from the stability analysis and in ground which is capable of supporting the anchor load.

The free length of tendon must be de-coupled from the surrounding grout so that it extends when stressed to the required load during cyclic testing and the final working load.

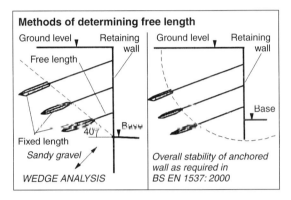

Methods of determining free length

Ground level — Retaining wall

Free length

Fixed length

Sandy gravel

40°

Base

WEDGE ANALYSIS

Ground level — Retaining wall

Base

Overall stability of anchored wall as required in BS EN 1537: 2000

Anchor spacing both vertically and horizontally will be determined by the wall stability requirements – usually a minimum of 1 m centres horizontally or 4 times the largest diameter of the fixed anchor. Site geometry has to be considered.

Vertical spacing must conform to the structural design of the wall; it will also take account of the construction methods.

Anchor protection is provided by encapsulation of the tendon with cement or resin in corrugated plastic ducts over the tendon bond length. The ultimate bond stress between this encapsulation and the anchor grout is 3 kN/m^2 with a minimum length of 2 m. The free length of tendon should also be sheathed in ducting or heat-shrink sleeve which allows movement during extension of the tendon; a corrosion inhibitor is also recommended. It is preferable for the corrosion protection and end encapsulation to be made in factory conditions.

The anchor head transfers load from the anchor to the retaining wall. It consists of cylindrical mechanical wedges and collets for strand anchors and of a nut and thread for bar anchors, on a steel bearing plate. The bearing plate may be bedded into the concrete wall or be supported on a waling along the wall.

The anchor head will also require corrosion protection and may need to be designed for re-stressing. Load cells for monitoring may also be incorporated.

Construction methods must ensure that a stable hole is formed so that the anchor tendon can be inserted and grouted in accordance with the design, with minimum disturbance of the surrounding ground.

Drilling techniques are varied, but generally use simultaneous rotary percussive drilling and casing in granular soil, with water, air or foam flushing. Hollow-stem augers are useful in cohesive soils. The hole must suit the size of anchor being used – typically 2 times tendon diameter, say, 100–250 mm diameter. Alignment of the drilled hole must be accurate so that interference with adjacent holes does not occur, but some vertical deviation (droop) is likely with inclined holes. A borehole survey may be needed in congested anchor zones.

If loss of grout is likely over the anchor zone (potential for absorption of, say, 5–10 lugeons), then pre-grouting and re-drilling may be necessary.

The hole must be clear of debris before placing the anchor. This is particularly important in under-reamed or belled anchors.

Tendon cables using strands (to BS 5896: 1980) are best fabricated off site in factory conditions; bar tendons (to BS 4486: 1980) are supplied with threads and usually have the protection sheathing applied on site. Centralisers should be attached so that the tendon fits centrally in the hole. Tendons must be protected and stored flat on site and carefully inspected for damage, corrosion or contamination before use.

Grouting should be undertaken as soon as possible after the tendon is placed in the hole. Grout constituents should be weigh-batched and thoroughly mixed in colloidal mixers; injection should be continuous for each stage or lift.

Type A and type D anchors are grouted using a tremie pipe placed to the bottom of the hole.

Type B anchors may be grouted through the temporary casing as it is carefully withdrawn in lifts at pressures up to 10 bar.

Type C anchors require hydro-fracture pressures >20 bar; usually using TAM methods so that primary and secondary injections can be made.

Stressing is carried out by extending the tendon using calibrated hydraulic jacks for strand anchors and a torque wrench for bars, reacting against the correctly installed anchor head.

Stressing must wait until the grout has reached the required strength (>30 N/mm^2) and no tendon must be stressed >80 per cent of the characteristic strength. Proof loading of a number of anchors to 1.5 times working load will be specified and each anchor will be subject to an acceptance test, all as in BS 8081: 1989.

Where appropriate, re-stressable anchor heads can be used to compensate for movement of the retaining wall – e.g. consolidation of the soil under load. The lock-off load may need to allow for relaxation during consolidation.

It may be possible to post-grout anchors which fail during testing (say, TAM anchors) and re-test. If not, or if the anchor fails on re-test, then the anchor is usually rejected and a replacement installed. For temporary applications the anchor may be used at say 50 per cent of the test load achieved.

> **Caution**: personnel must not stand or pass in front of the jacking head during stressing, and barricades and warning notices of operation must be provided.

Monitoring and recording of all anchor installation and performance procedures should be as in BS 8081: 1989; any variations in conditions encountered compared with the design assumptions should be reported immediately to the designer for assessment. Movement of the retaining wall and adjacent buildings during excavation, drilling and stressing must be checked – with electro-levels and inclinometers. Access to read load cells in anchor heads may be difficult.

9.2 Holding-down Anchors

The principles of holding-down anchors are similar to those described above for retaining wall tie-backs, but methods of design and construction vary.

The anchors may be tensioned to limit movement of the structure as above or act as 'dead' or 'passive' anchors.

Suitable for resisting vertical, horizontal and fluctuating loads due to:

- upward movement of buoyant foundations and structures in high and variable groundwater conditions
- wind loads on towers and pylons causing overturning
- bridge suspension cables
- strengthening and raising the height of concrete dams
- horizontal berthing forces for mooring dolphins.

Anchors are a cost-effective alternative to tension piles in these conditions, particularly where inclined tension resistance has to be provided.

Usually installed in rock, but may be appropriate in soils where adequate weight of soil can be mobilised; they can be used for stabilising soil slopes (see also 'Soil nailing', 9.4). High-capacity anchors (up to 1,000 tonnes) are now being installed for dam repairs requiring special attention to long-term creep (Bruce, 1997).

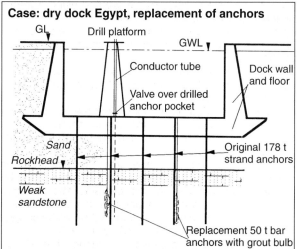

Case: dry dock Egypt, replacement of anchors

Anchor wires in the original 178 t anchors were corroding beneath the bearing plate.
Depending on the lift-off test carried out at each anchor, up to 4 No 50 t replacement anchors were installed in the weak sandstone around suspect anchors.
Groundwater level was approximately 13 m above dock floor, which meant that all drilling and anchor installation had to be carried out from a drilling platform above this level, as shown.

Ground investigation is similar to that for tie-back anchors. Fissures, infilled joints and RQD will be important in deciding the depth of anchor and the bond length in rock. Where anchors in soil and weak rock are to be subjected to transitory or repeated loading, cyclic loading tests on trial anchors may be needed to check design assumptions.

Design methods are described in BS 8081: 1989. Attention is drawn to potential loss of capacity for anchors in soil undergoing repeated loading.

The pre-stressing load applied to this form of ground anchor needs careful consideration – e.g. the requirement for the anchor to resist buoyancy of an empty dry dock but also to cope with applied load from vessels.

If sound rock is found a reasonable distance below the foundation it is usually preferable to place the anchor in the rock rather than in the overburden.

For tensioned anchors design is similar to type A anchors in BS 8081: 1989 specifically:

- bond stress between anchor and grout, and grout and rock
- allowable stress in the tendon
- weight of soil and rock which is lifted by the anchor (see Tomlinson, 1994, for charts to estimate appropriate resistance volumes).

For dead anchors, which may be bar tendons or larger-diameter steel tubes, the same design principles apply.

When tension piles are not pre-stressed by ground anchors a large factor of safety has to be applied to the shaft's frictional resistance.

Bond zones are restricted by some authorities to below the top 3 m of rockhead. Pre-grouting of fissured rock may overcome such a limitation; pre-grouting of karstic limestone to produce the bond zone requires considerable care to flush out soft infill in fissures.

Anchors in weak rock such as weathered chalk must take account of the reduced grout–rock bond available due to softening during drilling (Lord et al., 2002).

A compression plate or nut at the end of the tendon may be preferable to increasing the bond length where allowable grout–rock bond stress is low.

Corrosion protection should be provided over the whole length of the tendon, with double protection (in addition to the encasing grout) over the tendon bond length.

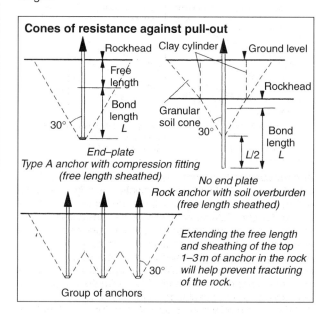

Cones of resistance against pull-out

Type A anchor with compression fitting (free length sheathed)

Rock anchor with soil overburden (free length sheathed)

Extending the free length and sheathing of the top 1–3 m of anchor in the rock will help prevent fracturing of the rock.

Group of anchors

Construction of type A and dead anchors in rock requires a stable hole, usually pre-grouted.

Anchor tendons using cables or bars should be prepared off site wherever possible to ensure quality. Larger tubes for dead anchors will usually require onsite welding of shear connectors

Drilling may be by rotary, rotary percussive or coring methods using air, water or foam flush; down-the-hole hammer drills are particularly effective in sound rock.

Hole diameter should be around 2 times anchor diameter for type A, but for larger dead anchors the annulus around the anchor tube should be limited to 50 mm to maintain the steel–grout bond.

The anchor should be inserted as soon as possible after-drilling and cleaning the hole. Where multiple long (50–80 m), highly stressed strand anchors are placed close together – e.g. for strengthening or heightening a concrete dam – drilling must be accurate to at least 1 in 100 (Shields, 1997).

Grouting is usually carried out in one or two stages through a small-diameter plastic tremie pipe placed to the bottom of the hole, or through hollow bar anchors.

Dead anchors may be constructed from the drill tube used to make the hole, and grouted through the bit 'lost' at the end of the tube.

Large tubular steel anchors will be inserted into a stable drill hole and the annulus grouted with a steel grout tube temporarily locked into a special grout shoe. Care is needed to prevent flotation of the anchor tube.

Low water:cement ratio grout (0.4–0.5) should be mixed in high-shear mixers to reduce bleeding.

Steel fibres have been added to grout to improve the tensile strength of temporary anchor grout.

Stressing jacks on multiple strand anchor tendon

Stressing, as specified, is carried out by extending the free length of the tendon using jacks or a torque wrench. For multi-strand, highly stressed anchors it is usual to apply an initial load to each strand (say, 5 per cent of the working load) using a mono-jack before stressing the whole anchor to working load with a multi-jack to ensure that all strands take an equal load.

Proof loading (up to 1.5 times working load; 80 per cent of ultimate tendon strength with appropriate cycling) of 10 per cent of anchors and acceptance tests for each anchor are normally specified. Long anchors may experience frictional resistance from the sides of the borehole if deviation and corkscrewing exists in the hole.

> **Caution:** the requirements mentioned in 9.1 concerning safety during stressing of anchors must be observed for all anchoring operations.

Monitoring should include the elements covered in 9.1 for tie-back anchors:

- drilling position and alignment – particularly for close groups
- water testing, pre-grouting and cleaning of the hole
- tendon preparation and corrosion protection
- tendon grouting and grout strength
- tendon stressing and testing
- movement of the structure.

OTHER TYPES OF GROUND ANCHOR

Single-bore multiple anchors

The progressive de-bonding of the grout–tendon or grout–ground interface will produce non-uniform stress along the fixed length. In order to improve the stress distribution in the ground, the concept of the 'single-bore

multiple anchor' has been patented (Barley, 1997). This provides simultaneous loading of a number of short, fixed-length anchors in a single hole. Each tendon strand has its own fixed length of tendon staggered in the hole. It is stressed with its own mono-jack, but the whole anchor is stressed simultaneously by synchronising the jacks.

Safe working loads of 2,000 kN for tie-backs in soils are reported by Barley and McBarron (1997) and Jewell (2003).

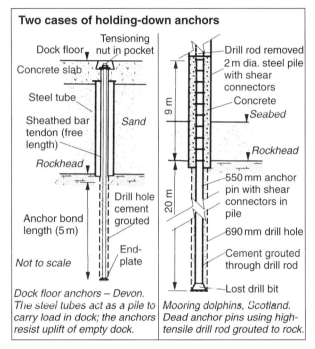

Two cases of holding-down anchors

Dock floor anchors – Devon. The steel tubes act as a pile to carry load in dock; the anchors resist uplift of empty dock.

Mooring dolphins, Scotland. Dead anchor pins using high-tensile drill rod grouted to rock.

Removable ground anchors

Tie-back anchors are a useful alternative to large and costly internal strutting of retaining walls (Herbst, 1997). However, legal limitations on installing pre-stressed anchors under adjacent structures and the need to accommodate future underground services and excavations in anchor zones have led to the development of anchors which can be safely removed after providing temporary support. The removable temporary anchor system must:

- conform to the design requirements of BS 8081: 1989
- be replaced by other structural support within the permanent building before removal.

The simplest form of removable anchor is the threaded bar anchor, which can be back-threaded off the short fixed anchor length which remains unstressed in the soil. Strand tendons are difficult to de-couple; over-stressing of a weakened section of the tendon above the anchor zone may be practical and electrical heating of the strands has been attempted.

The single-bore multiple anchor can provide a patented means of de-bonding (Barley and McBarron, 1997).

Use of non-metallic tendons

Carbon fibre rods are now used for permanent soil nailing (9.4) because of high tensile strength (a 12.5 mm diameter rod has a breaking load of 250 kN), good corrosion resistance and good durability. They are also light, and easy to handle and cut to length. Application to tensioned ground anchors has not been significant.

Jet grouted anchors

Techniques to insert bar anchors into jet grouted bulbs in sands and gravels to form the anchor zone have not gained much support to date.

9.3 Rock and Cliff Stabilisation

The stabilisation of rock faces and slopes utilises a wide variety of techniques, many stemming from long-established mining methods such as rock bolting, rock anchors, dowels and blasting.

To ensure safety, the design and construction of stabilisation works for new and existing slopes, large rock excavations and underground caverns require a sound knowledge of geological structure, rock mechanics and hydrogeology.

The principle of anchorage for stabilisation is to form a reinforced zone in the rock mass which minimises the movement along discontinuities within the mass.

Suitable for controlling instability in natural and excavated slopes caused by:

- the slope being too high or too steep for the materials making up the slope
- weak materials and physical discontinuties
- rock falls
- water pressure and seismic forces.

Ground investigation should follow the basic principles of a phased programme to produce the conceptual ground model. The objective should be the classification of potential slope failure and movement, particularly the dip, strike and frequencies of bedding and joint systems. The desk study will reveal areas where more specific data are needed for the ground model, e.g. from air photography, surface geological mapping, horizontal photogrammetry, trial pits and core drilling in order to make preliminary assessments of slope stability.

This will be relatively straightforward for exposed slopes, where access can be obtained either on mobile platforms or by abseiling techniques, but more difficult when reliance has to be placed on interpreting discontinuities and orientation of core samples drilled from the surface behind a proposed slope.

In addition the investigation may need to determine the most appropriate excavation method – direct, ripping or blasting – by using rock mass classification systems and other data (1.4).

The presence of groundwater in the slope will have a significant influence on stability, and, as seen for excavations in soil, drainage can be the most effective means of improving the stability.

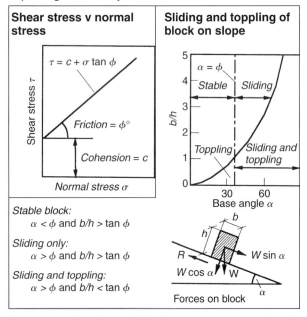

Shear stress v normal stress

$\tau = c + \sigma \tan \phi$

Shear stress τ

Friction = $\phi°$

Cohension = c

Normal stress σ

Sliding and toppling of block on slope

$\alpha = \phi$

Stable | Sliding

b/h

Toppling | Sliding and toppling

Base angle α

Forces on block

R $W \sin \alpha$

$W \cos \alpha$ W

Stable block:
$\alpha < \phi$ and $b/h > \tan \phi$

Sliding only:
$\alpha > \phi$ and $b/h > \tan \phi$

Sliding and toppling:
$\alpha > \phi$ and $b/h < \tan \phi$

Design of effective stabilisation measures requires the analysis of the mechanics of slope failure (Bromhead, 1992). Once this is established, BS 8081: 1989 provides information on the use of rock bolts and anchors to design the appropriate stabilisation works.

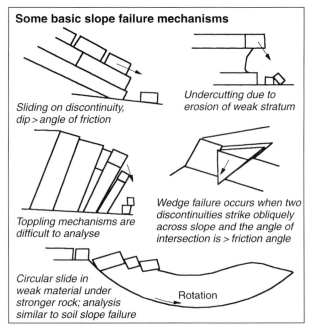

Some basic slope failure mechanisms

Sliding on discontinuity, dip > angle of friction

Undercutting due to erosion of weak stratum

Toppling mechanisms are difficult to analyse

Wedge failure occurs when two discontinuities strike obliquely across slope and the angle of intersection is > friction angle

Circular slide in weak material under stronger rock; analysis similar to soil slope failure

Rotation

The basic methods of stabilising an unsafe slope are:

- to cut it back to a flatter angle or reduce the height
- to construct a structural retaining wall
- to add weight at the toe.

Where these methods are not feasible due to lack of space, poor access, height and excessive cost, then the existing rock profile has to be secured *in situ*.

Using the basic relationship between shear stress and normal stress, the force to resist sliding of blocks (R) on an inclined plane can be determined:

$$R = cA + (W \cos \alpha) \tan \phi,$$

where A is the base area of the block and other terms are as in the adjacent diagram.

If porewater pressure (u) exists then total and effective stress conditions must be examined.

In order to improve the resistance, rock bolts (tensioned to a load of T at an angle of β to the slope) are used to increase the normal force and reduce the disturbing force on the block:

$$R - T \cos \beta = cA + (W \cos \alpha + T \sin \beta) \tan \phi.$$

Similarly, unstressed dowels may be used as shear connectors to increase resistance to sliding. Grouting of open joints to bond rock into a more stable mass may be needed to form the anchor zone.

More complex instabilities such as toppling, ravelling, and undercutting are not easily analysed and the design of remedial works relies on the experience of the geologist.

Stereograms (1.4) are useful graphical representations to show likely failure modes in rock slopes.

Observational methods should be applied to newly cut slopes to ensure that data revealed by the exposure of fresh rock are fully accounted for.

Photogrammetry techniques provide excellent data for the geologist to design the works and place the field information in context for the report. Where scaffolding covers the face to carry out the works, the prior photogrammetric views provide the construction crew with details of the work needed stage by stage.

Factors of safety are considered in two phases: that for the equilibrium of the rock mass, usually 1.5; and that for the pull-out of anchors, 1.5–2.5, depending on the service life of the anchor.

Rock anchors of type A (9.1) are used for high working loads (500 kN) with deep anchor zones (10–40 m); strands may be used as tendons for downward-inclined anchors, but for upward-sloping anchors bars will be used for ease of placement and grouting.

The fixed length is based on an ultimate bond stress of 10 per cent of the compressive strength of strong rock to a maximum of 4 N/mm² using >30 N/mm² UCS grout. In weak rock the bond stress should be related to shear tests on the rock. The recommended fixed length is 2 m in strong rock to 10 m maximum. Depending on the near-surface condition of the rock, the anchor may be fully bonded or be formed with a free length (necessary for re-stressing).

Stressing and cyclic loading tests are as described for tie-back anchors.

Double corrosion protection is required for permanent anchors (>2 years).

Rock bolts using mechanical fixing or polyester resin capsules (9.5) over the bond length with solid or hollow bars (25 mm) are used in hard rock, care being taken to match the diameter of the hole to the size of the end fitting. Once the anchor is stressed (60–100 kN) it is usual to grout the shaft fully. Manufacturers' data on deformed bars and bonding material provide estimates of pull-out resistance.

Dowels (un-tensioned bars) are used to pin rocks near the face, to provide shear connectors to new concrete underpinning and existing rock. They are suitable for most rock types, using bond values as for anchors. Corrosion protection wrapping is usually applied on site for bolts and dowels.

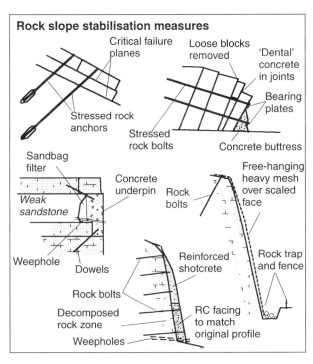

Rock slope stabilisation measures

Critical failure planes
Loose blocks removed
'Dental' concrete in joints
Bearing plates
Stressed rock anchors
Stressed rock bolts
Concrete buttress
Sandbag filter
Concrete underpin
Rock bolts
Free-hanging heavy mesh over scaled face
Weak sandstone
Weephole
Dowels
Reinforced shotcrete
Rock trap and fence
Rock bolts
Decomposed rock zone
RC facing to match original profile
Weepholes

Underpinning with reinforced concrete is used to support overhanging blocks when removal is not possible. Access for shuttering and placing concrete is needed, usually from scaffolding. The underpin itself is either doweled or bolted to the rock prior to the block of rock itself being bolted.

Concrete 'dentition' is the term applied to raking out large infilled joints at the rock surface and ramming in concrete or mortar to prevent water ingress and root growth. Care is needed to ensure that the blocks around the joints being treated do not loosen during the work.

Cable restraints are used as temporary means of holding unstable blocks in place so that underpinning or

bolting of the blocks can be safely carried out. The cables are fixed to anchor points drilled and grouted a safe distance from the block.

Fissure grouting and pressure pointing using cement grout and mortar to prevent water entering joints and improve bonding of the rock mass are carried out in conjunction with rock bolting. Pre-grouting will be necessary where loss of anchor grout is likely.

Epoxy resin has also been injected successfully into fine fissures where cement would not penetrate at acceptable pressures and where improved bonding was required. Ultimate bond strengths of over 30 N/mm² have been measured in compression across a 45° joint in basalt (Price and Plaisted, 1971). Epoxy resin can also be used for injecting around anchors and for surface coating of rock to reduce water penetration – some discoloration may occur with time.

Caution: onsite mixing of epoxy resins demands a high degree of accuracy in batching materials to maintain quality.

COSSH Regulations apply (HSE, 2002): materials must be handled with care and protective clothing and eye protection are necessary. Solvents are flammable.

Protection netting is an effective means of controlling rock falls after thorough scaling down of loose material. Chain-link mesh is generally used, but where larger blocks may be loosened by weathering, heavy-duty netting is needed in addition. The nets may be free hanging or bolted so that a space is left behind to allow rocks to fall and not be trapped at a high level.

Rock traps and fences are usually provided at the base of cliffs protected by netting.

Abseiling to install rock anchors

Construction requires access to the rock face. Where considerable works are required involving detailed surveys, concreting, grouting, bolting and highly stressed anchors it is usually necessary to erect steel scaffolding over the area to be treated; this adds considerably to the cost. Long-reach hydraulic excavators fitted with drill masts are effective where access to the base of the cliff is available and the bolts can be driven and remotely fixed.

For more straightforward works – e.g. to scale down the face, insert short bolts

Netting to rock face

and place protective netting – abseiling techniques are now well developed.

Monitoring will start with the instrumentation for determining the stability of the rock slope for design calculations, e.g. measuring movement of tension cracks and rock mass, groundwater levels. Once established in accessible locations these devices can be used to check the long-term stability of the rock face. Load cells may be installed at the head of rock bolts and anchors.

9.4 Soil Nailing

Soil nailing is an adaptable form of slope stabilisation in soil and has developed significantly, first as a means of providing temporary support during construction and recently for permanent retaining structures and wall and slope repairs.

The principle is to provide tensile and shear resistance within the soil by drilling or driving in passive (untensioned) steel or carbon fibre bars in a close grid to form a stable gravity structure. The nails act as a group and remain unstressed until ground movement occurs, at which time the shear, tensile and bending resistance of the nails is activated. A hybrid system that nails and pre-stresses the soil is under development (Jewell, 2003).

Suitable for reinforcement of existing and newly cut slopes in most types of cohesive and granular soils; also applicable to unstable soil retaining walls. Tensioned ground anchors (9.1) may be used for slope reinforcement in conjunction with soil nails.

Not recommended in soils with no cohesion or high clay content, in caving sands or below the water table.

Ground investigation for remedial works to a failed slope must determine the location and geometry of the failure planes, together with residual shear strength parameters, soil density and particle size, the water table and angle of internal friction, as required for new cuttings. Data should be acquired for a distance of 3 times the slope height from the crest of the slope and for every 200–600 m² of slope face.

A test programme to check pull-out resistance of trial nail lengths will assist in economic design.

Design guidance is given in BS 8006: 1995 and in HA 68/94 (1994), based on analysis of external and internal stability at each stage of the excavation. The nails must provide pull-out resistance to the thrust of the failing wedge of soil – single or two-part wedge (modified Coulomb analysis) or a log-spiral wedge. The principle is to ensure that the force available from the total tension developed in the nails is greater than the force causing the wedge to fail. Typically, a factor of 1.5–2 is applied for permanent nails.

Failure mechanisms and analyses

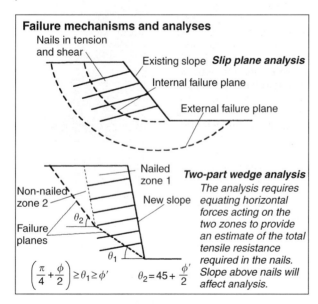

Nails in tension and shear

Existing slope *Slip plane analysis*

Internal failure plane

External failure plane

Nailed zone 1 *Two-part wedge analysis*

Non-nailed zone 2

New slope *The analysis requires equating horizontal forces acting on the two zones to provide an estimate of the total tensile resistance required in the nails. Slope above nails will affect analysis.*

Failure planes

θ_2

θ_1

$$\left(\frac{\pi}{4}+\frac{\phi}{2}\right) \geq \theta_1 \geq \phi' \qquad \theta_2 = 45 + \frac{\phi'}{2}$$

Wedge analyses require a number of iterative computer calculations using a range of trial nail spacing, inclinations and lengths based on experience of the particular soil parameters and wall height (Schlosser *et al.*, 1992).

The extensive French research programme RDGC – 'Clouterre' (soil nails) (1993) gives empirical guidance on the tensile and shear resistance mechanisms of the nail to be applied for stability. Also examined are the potential for stiff nails to break in bending and of the soil to fail in bearing below the nail. Estimates of horizontal and vertical displacements of the gravity block are given – ranging from 0.1 to 0.3 per cent of the wall height. The design methods are based largely on pressuremeter testing.

It is assumed that the nails form a retaining or gravity block, which also has to be checked for overturning and sliding due to the active pressure from soil behind the block and bearing pressure below the block. The stiffness of the nails will modify active earth pressure acting on the wall facing as will the slope of ground above the nailed wall.

Failure of the nailed slope may be as a result of:

- nails breaking
- lack of nail-to-soil adhesion
- excessive height of an excavation stage.

Slip circle analysis is used for overall stability of nailed retaining walls and shallow slopes using a potential failure surface which passes behind and beneath the gravity block, with due allowance for pore pressure.

Surcharge load on the nailed block and effects of existing structures at the top of the nailed wall must be examined.

Nails are usually steel bars of 20–50 mm diameter, 16 mm carbon fibre rods or glass fibre rods, depending on the tension resistance required.

Typical dimensions of grouted steel nails for a wall height H and $\phi = 35–45°$ are given in the table.

Length (L) of nail in granular soil	0.4–0.7H
Length (L) of nail in soft clay soil	1–1.2H
No. of nails per m²	0.5–2
Diameter of hole (mm)	75–100
Tensile strength of steel nail (kN)	120–200
Cumulative length of nails (m/m² face)	10–15

Frictional resistance to pull-out of drilled and grouted steel nails can be as high as 0.6 N/mm² in dense granular soils, but more typically ranges from 0.05 N/mm² in clays to 0.25 N/mm² in sandy gravel.

Carbon fibre reinforced polymer (CFRP) bars have high tensile strength (450 kN breaking load for a 16 mm diameter bar) but a low Young's modulus. Long serviceability (120 years) is claimed by manufacturers. CFRP is also resistant to environmental attack and no secondary protection is required. The bars are surface treated during manufacture to improve the bar–grout bond (0.1–0.2 N/mm²). They cannot be coupled on site, but are easily cut to length and installation rates are high.

Glass fibre rods may degrade when surrounded by alkaline cement grout and are therefore not recommended for long service.

Displacement of the gravity block horizontally is usually less than the vertical displacement by a factor of L/H and depends on:

- the rate of construction
- the bearing capacity of the soil
- the inclination of the nails
- the extensibility and bending stiffness of the nails.

Displacements are usually at a peak during excavation. Horizontal displacement at the top of wall varies from 0.001H in weathered rock to 0.003H in clayey soils.

Corrosion protection is required for all permanent applications. The cement-grouted bar within a plastic sleeve should give a service life in excess of 30 years. Where unprotected steel nails are driven into soil, an expendable or sacrificial allowance for corrosion should be made to extend service life. Stainless steel may be considered.

Facing is required to ensure stability of the slope between the nails; shotcrete thickness, reinforcement and bearing plate requirements will depend on the tension developed in the nails. The 'Clouterre' tests provide some end load versus displacement indicators.

Shotcrete is applied in most retaining wall applications as soon as possible after nailing a stage – but it may be necessary to apply shotcrete to stabilise the excavated face before installing the nails.

An alternative to shotcrete is a geotextile mat, which can be seeded to produce a natural-looking (relatively flat) slope with vegetation roots helping stability. These 'geomats', with a central pre-seeded coir blanket (to protect against wash-out), can be fixed to the nail heads where tensile forces at the face are low. Additional pins may be needed to keep the mat in place and stronger matting at the base of the slope.

Slope drains will usually be required at the toe of the wall and extending beyond the nails to ensure that water cannot build up behind the gravity block and facing. Drainage holes at 2–3 m centres are usually lined with slotted PVC pipe wrapped in filter fabric, with the discharge at the face directed away from the soil.

Lined interceptor drains may also be needed at the top of the slope to prevent surface water penetration; a drainage blanket behind the facing may deal with run-off.

It may be necessary to designate a specific sequence of drilling on the slope rather than simply follow the row of nails. In all cases care must be taken not to loosen the soil during installation.

Drilling and grouting methods are similar to rock bolting; casing may be necessary and pressure grouting applied as the casing is withdrawn over the nail. The method uses 50 mm nails in 100 mm holes and grout pressures <5 bar.

Alternatively, as the nail is usually inclined downwards, an open drilled hole can be filled with grout (by tremie pipe) and a plastic protection tube centred in the hole followed by insertion of the nail with centralisers.

'Self-drilling' nails are hollow steel bars (usually continuously threaded) with sacrificial bits which are drilled into the soil without casing. The hole may be simultaneously drilled and grouted with a low water:cement grout flush (0.5–0.7) through the hollow stem to improve the bond and effective diameter of the nail.

Driving of the nail using a conventional hydraulic hammer drill adapted to take the nail can install temporary nails quickly and cheaply – the method lacks adequate protection from grout or sheathing (ripped off as the nail is driven).

'Ballistic' shooting using a 'launcher' of a 38 mm diameter steel nail up to 6 m long into soft soil using compressed air has been used in appropriate conditions. Safety precautions include computer control of loading and firing (Myles, 1995).

Shotcreting requires experienced 'nozzle-men' to avoid erosion of the soil face.

Pull-out tests are recommended at a rate of five per 1,000 m², with one test for each excavation stage.

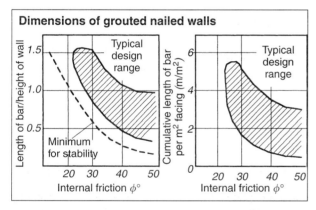

Dimensions of grouted nailed walls

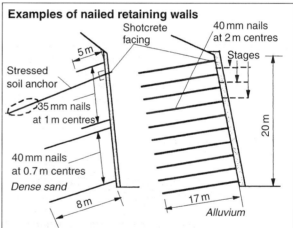

Examples of nailed retaining walls

Wall is excavated in stages so that nails can be drilled and grouted at each level. Shotcrete is sprayed over reinforcing mesh for each level

Construction sequence for new excavations, working from the top down, is to:

- excavate to the level of the first row of nails – usually a 1–2 m stage, but no greater than allowable for local stability of the slope
- install nails at that level as soon as possible after exposing the face
- apply reinforcing mesh and face coating to the excavated face
- repeat operations in stages until the base of the wall is reached
- form weep holes.

Stabilising existing soil slopes may be carried out using:

- light track/wagon drills working on the slope
- hydraulic arms with drill feeds operating from the top or bottom of the slope
- abseiling
- scaffolding platforms.

Monitoring of deformation of a high face during excavation will indicate whether additional nails or a revised installation sequence are necessary based on observations. Monitoring of tension forces in the nails and displacement of the slope during the service life is desirable, but access may be difficult.

Nailed reinforced earth

9.5 Underground Support

Excavation for tunnels and large underground caverns in rock may require a variety of support measures, depending on the rock condition, potential modes of failure and method of excavation – drilling and blasting, TBM, etc. Concrete lining, steel ribs, rock grouting and drainage, and NATM methods are typical; rock bolting methods are used to form a stable arch or beam in the rock mass around the opening and to restrain sliding blocks.

As in cliff stabilisation, support is required to prevent failure of the rock mass due to:

- movement of blocks of rock into the excavation due to *in situ* stress or external forces such as water pressure and gravity
- over-stressing of intact rock.

Ground investigation will need to cover all the issues considered previously for rock excavations (1.4) and for cliff stabilisation (9.3). Some additional parameters are:

- *in situ* stress measurements using stress meters and strain cells in boreholes
- deformation modulus from a pressuremeter
- classification of the rock mass rating (RMR in 1.4) or the Norwegian quality index (*Q*) (Hoek and Brown, 1980) for stand-up times and empirical reinforcement
- swelling potential
- seismic activity and effects of blasting.

Risk assessment for underground excavations must be comprehensive and will have to determine costs of overcoming difficulties such as:

- large inflows of water
- faults, fractures and the associated over-break potential
- buried geological features such as infilled channels and irregular rockhead.

Even with high initial expenditure on a comprehensive investigation (say 3 per cent of the project cost), many of these features will only be revealed during construction. Risk assessment should provide observational methods and remedial plans to deal with them if and when they arise.

Rock bolts and dowels

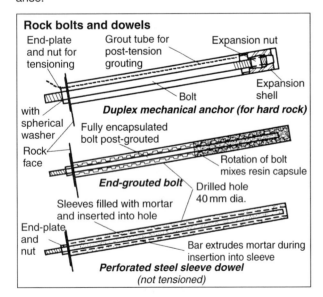

End-plate and nut for tensioning

Grout tube for post-tension grouting

Expansion nut

Expansion shell

Bolt

Duplex mechanical anchor (for hard rock)

with spherical washer

Fully encapsulated bolt post-grouted

Rock face

Rotation of bolt mixes resin capsule

End-grouted bolt

Drilled hole 40 mm dia.

Sleeves filled with mortar and inserted into hole

End-plate and nut

Bar extrudes mortar during insertion into sleeve

Perforated steel sleeve dowel
(not tensioned)

Design methods include analysis of stability and stresses using powerful computer programs and, to a lesser degree, empirical relationships (Douglas and Arthur, 1983).

Preliminary design will assess potential failure mechanisms based on simple homogeneous conditions and identify possible means of support and the data needed for detailed design. Finite element analysis then allows complex geometries and variations in rock strength and properties to be considered in 3D.

Structural stability considerations, particularly in respect of wedges formed by discontinuities in the rock mass, in conjunction with the analysis of stress distribution, determine the reinforcement necessary. For shallow excavations stability depends largely on structural parameters and degree of weathering of the rock mass. Stereograms are used to indicate the volume and direction of movement of potentially unstable blocks in the rock mass.

Rock reinforcement alone is unlikely to be appropriate if:

- support pressure >600 kN/m^2
- spacing of dominant discontinuities >600 mm
- the rock strength is inadequate for anchorages
- RQD is low or there are infilled joints, high water flow.

In such cases external support using steel ribs will be indicated.

Post-reinforcement to support blocks

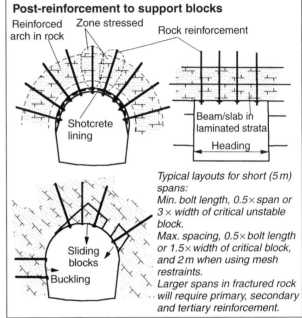

Reinforced arch in rock

Zone stressed

Rock reinforcement

Shotcrete lining

Beam/slab in laminated strata

Heading

Sliding blocks

Buckling

Typical layouts for short (5 m) spans:
Min. bolt length, 0.5× span or 3× width of critical unstable block.
Max. spacing, 0.5× bolt length or 1.5× width of critical block, and 2 m when using mesh restraints.
Larger spans in fractured rock will require primary, secondary and tertiary reinforcement.

Rock anchors of type A (9.1) are suitable for reinforcing most rock – mainly single bar for ease of installation in confined spaces, but if needed higher ultimate capacity can be provided by strands. Double corrosion protection is required for permanent anchors and proof loading of each anchor during tensioning is normal.

Rock bolts with mechanical fixings are suitable for hard rock, but grouted fixed-length bolts (resin capsules or cement) can be used in most rocks. Length varies from 2 to 8 m for a resin capsule grouted bar to 3–20 m for an expanding shell fixing on a bar. Protection is by prewrapping the free length and grouting after tensioning. Upward-inclined bolts have to be cement-grouted through end plates.

Dowels have a variety of uses in most rock types:

- pre-reinforcement prior to final profile excavation
- strengthening rock near to the surface of the excavation while anchors are installed
- support for reinforcement mesh for shotcrete lining.

Dowels may be deformed steel bars, glass fibre or plastics, depending on whether they are permanent or to be removed during excavation. To avoid the need for packer grouting of dowels, various proprietary methods have been developed, e.g. a 'split steel' perforated tube filled with mortar into which the steel bar is rotated. Resin capsules are also used extensively.

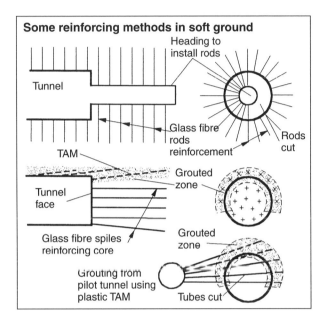

Case: pre-reinforcement using forepoling for tunnel enlargement, Switzerland

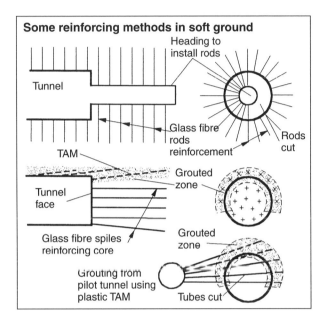

Section

The steel forepoles were drilled to an accuracy of 1 in 200 to ensure underground services were avoided. Sealed grout ports at 1 m centres in the tubes were ruptured at low pressure and a fixed quantity of cement grout injected to form a strong umbrella over the excavation for the enlarged tunnel.

Spiles or dowels which can be cut through during excavation are used to reinforce full-face cutting of a tunnel in clayey soil or weak rock. The typical sequence of work is:

- reinforcement of the face with close-centred grouted glass fibre rods (the spiles)
- mechanical excavation of the face and spiles for a safe distance, leaving adequate protection ahead
- placement of arch rings
- shotcreting of the new face and arch rings
- repeating the reinforcement process, etc.

Construction by open excavation under ground – as opposed to TBM methods – can require a large amount of hole drilling, which can be undertaken with a variety of rotary percussive drilling tools from small 'stopers' to jumbo rigs capable of drilling several holes concurrently. Holes are required for blasting, rock bolts, support for reinforcement mesh, pre-grouting, back grouting and drainage. Special rigs have been built for drilling and installing resin capsule rock bolts – particularly useful for roof bolting in tunnels with high crowns.

Installation and grouting of whatever type of rock reinforcement is used should be completed as soon as possible after excavation of the tunnel section.

Blasting is a highly specialised technique and outside the scope of this *Introduction*.

Installing spiles at shotcrete tunnel face

Shotcrete (sprayed concrete lining) is used extensively for temporary or primary support, apart from the designated NATM method. Where temporary steel arches are needed, the ribs and the length of tunnel between the ribs are usually treated with reinforced shotcrete. The thickness of the shotcrete will be from 300 mm (in two layers) for rock with RMR < 20, to 50 mm for RMR >80. Thickness must be controlled by working to steel pegs set out to the design requirements – most important in NATM.

Reinforcement of shotcrete with 3–5 per cent by weight of 50 mm steel fibres as an alternative to mesh reinforcement produces a strong lining, but attention must be paid to mixing, pumping and nozzle equipment to avoid blockages.

Some reinforcing methods in soft ground

Jet grouting to form an aureole or umbrella over soft ground tunnels (as with injection grouting to exclude water in 5.8) can provide a strong stable substrate for application of a shotcrete or structural concrete lining. The replacement of *in situ* soil with cement grout has to be limited, otherwise the disposal of spoil can be time-consuming. Difficulties will occur when attempting to use jet grouting where groundwater exists – the need to use stuffing boxes to counteract pressure and the need to allow surplus spoil to escape cannot be reconciled.

Monitoring the behaviour of the underground opening and surrounding rock mass is essential during construction, particularly as the observational method will be used to some degree during and after construction to demonstrate stability.

Instrumentation for rock stress, displacement of rock mass, lining and reinforcement, water pressure, anchor load, etc. should be installed as soon as possible. Measurement for soft ground tunnelling (Clayton *et al.*, 2000) will include:

- performance of the lining using embedded stress cells
- convergence using 'total station and targets' accurate to 1 mm (typical limits allowed in caverns and tunnels range from 0.1 to 1 per cent of width respectively)
- surface movement
- movements in buildings with electro-levels.

Shotcrete lining quality can be checked with coring and slot cutting, but this should be limited as disruption to the lining may cause problems.

10 PILE GROUTING

10.1 Piles to Offshore Oil Platforms

The large steel and concrete platforms installed in the North Sea and around the world in offshore oil fields have to be fixed to the seabed to support high dead and live loads and to resist fluctuating wave forces.

The foundations for most steel platforms (jacket structures) are piled into the seabed through sleeves connected to the legs, and the tubular piles are fixed to the platform with high-strength cement grout in the annulus between pile and sleeve.

Concrete platforms rely on the weight of the structure for stability, with the skirt around the large gravity base penetrating the seabed and the void between the underside of the base and the seabed grouted.

The principle of pile grouting is to bond the pile to the jacket using grout which will have high shear strength but will not shrink away from the sleeve or pile. In order to reduce the bond length necessary the piles and sleeves have shear keys welded at close centres.

Ground investigation for offshore piles and for bearing capacity for gravity platforms, whilst beyond the scope of this book, must be deep enough to examine the stressed zone (>100 m below the seabed in places) and to determine driving and frictional resistance characteristics for partially penetrating piles.

Design of offshore installations is covered in Regulations from the TSO (1996b). The ultimate bond strength is based on empirical formulae developed from extensive laboratory testing of piles up to 2.2 m diameter (Billington and Tebbett, 1980). It is dependent on the characteristic UCS of the grout, the stiffness of the combined pile and sleeve and a coefficient depending on the ratio of grouted length to pile diameter. A safety factor of 6 is used between the applied bond stress and the ultimate bond strength.

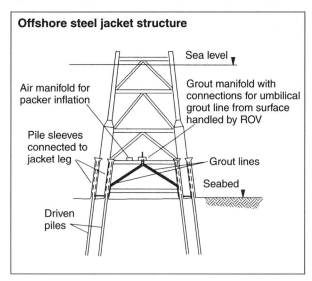

Offshore steel jacket structure

Sea level

Air manifold for packer inflation

Grout manifold with connections for umbilical grout line from surface handled by ROV

Pile sleeves connected to jacket leg

Grout lines

Seabed

Driven piles

Grout strength to produce the designed bonding between pile and sleeve requires high characteristic unconfined compressive strength from an early age. Cements used are generally rapid-hardening OP cement (RHOPC) or Class G API 'Oilwell' cement, depending on availability.

Low water:cement ratios of 0.35–0.4 with a fluid density of around 2 tonne/m^3 are necessary to meet these high UCS and shear strengths. Other benefits are that the amount of water which bleeds from the mix is limited and shrinkage is very low.

Provided that mixing produces a colloidal suspension and the pumping distance from mixer to sleeve is, say,

<200 m, the rheology of this neat cement mix should be acceptable for pumping without the need for plasticisers. Typical UCS at seabed temperatures of 8°C at 3 days and 28 days are, respectively:

- RHOPC   35 N/mm^2, 65 N/mm^2
- Oilwell G   30 N/mm^2, 60 N/mm^2.

Bond strength for an *L/D* ratio of 2 is taken as 2.5 per cent of UCS. These strengths are achieved with cements mixed with seawater.

As the UCS is directly proportional to the density, this is the main parameter used for quality control during grouting.

Where 'insert piles' are required below the driven pile to achieve adequate frictional resistance in weak soils, it may be necessary to lighten the grout with an additive (e.g. gilsonite) to prevent fracture of the strata, resulting in some reduction in bond strength.

Grout distribution pipes are a key issue in the design of the grouting system. For smaller platforms the pipes from each pile sleeve may be brought to the surface, but where a large number of primary and secondary (backup) lines are necessary, special remotely operated umbilical connections are used to deliver grout to each sleeve in turn.

Entry into the sleeve is at the lowest point, just above an inflatable packer or passive seal which closes the annulus after pile driving.

Installation of the piles is carried out as soon as possible after the platform is placed in position offshore; grouting follows on immediately afterwards. In certain sea areas where there is a high storm risk, pin piles are driven and grouted initially at the jacket corners to give early security. Pile 'grippers' may be used to hold these piles during grouting to give extra security at this stage.

Grouting plant is usually deployed on a construction barge anchored alongside the platform or held under dynamic positioning. The large amount of cement required may require additional deliveries from the shore base during the grouting operation.

Grout mixing must be rapid but thorough; high throughput rates for the low water:cement grout (up to 0.6 tonne/min) must be available for a large-volume annulus (10 m^3). Re-circulating jet mixers with data processing densimeters to control cement and water delivery continuously to ±0.5 per cent accuracy are now normal. Printouts of density with time should be provided for quality assurance.

Grout injection will commence after the packers (where provided) at the base of the annulus have been inflated and the annulus flushed clean. In the event of a seal failure it will be necessary to inject a small quantity of grout through the primary injection port to form a plug. When this is set, grouting of the remaining annulus continues through the secondary port above the plug.

Connections of the grout lines to each annulus may be made by divers to a central valved manifold at the top of the sleeves. However, it is more economical to use a remotely operated vehicle (ROV) to deploy an umbilical grout line to a special connector at each sleeve.

Injection continues until overflow occurs at the top of the sleeve, when a density check is made.

Pumps can be plunger pumps or positive displacement screw pumps capable of outputs of 1 m^3/min at pressures up to 50 bar (although grout will usually siphon through the annulus and to the overflow vent due to the height of the pumping station above the injection point). Higher output may be needed for initially flushing the annulus of long, large-diameter piles to give good cleaning velocity.

Monitoring of the density, and hence the compressive and bond strength of the grout during placement, is provided by the data controls at the mixer; flow and pressure recorders are provided at the pump. In addition manual density checks using a mud balance are made regularly.

The grout is checked at the top of the sleeve after 10–20 per cent over-run of grout has been delivered (to ensure that all water in the sleeve has been displaced) using a densimeter probe handled by the ROV which made the grout connections. The probe is placed in a docking device at the pile top and a direct readout of the density is provided to the engineer at the surface (Evans et al., 1980).

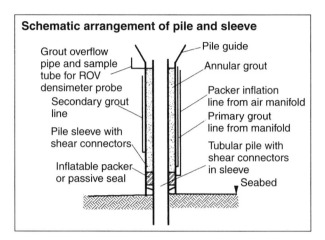

Schematic arrangement of pile and sleeve

Grout overflow pipe and sample tube for ROV densimeter probe
Secondary grout line
Pile sleeve with shear connectors
Inflatable packer or passive seal
Pile guide
Annular grout
Packer inflation line from air manifold
Primary grout line from manifold
Tubular pile with shear connectors in sleeve
Seabed

Test cubes (75 mm) of the grout are taken from the mixer at three stages during injection to coincide with the top, middle and bottom of the annulus. A testing laboratory is provided offshore to check the early strength of cubes.

UNDERBASE GROUTING

Gravity bases for offshore oil installations are large steel or concrete structures which can store oil in the integral caissons used as flotation tanks during the tow out to the offshore production site. Once on location, the tanks are flooded and ballasted as needed to place the platform on the seabed. Below the base of the caisson tanks are skirts which penetrate the soft seabed to provide lateral stability and reduce scour.

The principle of the underbase grouting is to fill any voids which may exist between the seabed and the underside of the caisson so that:

- settlement is reduced by improving stress transfer to the seabed
- soil is not removed by seawater entering the void.

Ground investigation will determine the safe bearing capacity of the seabed and the strata which will be stressed by the platform.

Design of the base will depend on the allowable bearing capacity, as in the relevant regulations (TSO, 1996b).

Grout strength and elastic modulus will be designed to be compatible with the soil strength – say, UCS of between 0.2 and 1.5 N/mm^2 at 28 days and density 1.2–1.4 tonne/m^3. Strong grout between the base and the soil may produce hard points and cause distorting stresses in the structure.

Mix proportions to meet these requirements must produce a grout made with seawater which:

- is cohesive, stable and will not shrink
- has low heat of hydration and low bleed
- has low viscosity but is capable of displacing seawater in the void.

A typical grout comprises a mix of OP cement with a high water:cement ratio, say, >2.5, and up to 14 per cent sodium silicate to control bleed to <2 per cent. It may be

useful to add PFA, ground granulated blast-furnace slag and/or bentonite to the mix in order to reduce the heat of hydration and improve rheology so that separation of the constituents does not occur (Domone, 1990). In view of the large volumes of material needed (possibly >20,000 m^3 for a void up to 2 m deep) these additives should be pre-blended on shore.

Grout distribution lines to the individual cells below the base require careful design to ensure that flow in the cell is uniform, scour does not occur and water is displaced through large vents. Each injection line will have a dedicated flow meter and control valve on a distribution manifold within the caissons.

Offshore grout plant

Grouting operations should commence as soon as the gravity base is ballasted on to the seabed to the design depth of penetration. As for pile grouting, the plant is deployed on an attendant construction barge, with injection pumps possibly located within a caisson above the cells.

The flow to each cell must be checked and the cell flushed prior to grouting.

Grout mixing and injection plant is similar in design and output to the pile grouting operation, but a pre-mix of the seawater and silicate is required before adding the cement. The injection rate will be low to start with, increasing to a maximum of around 1 m^3/min.

Injection is continued until the overflow grout is shown to be at the designed density for a period of around 15 min.

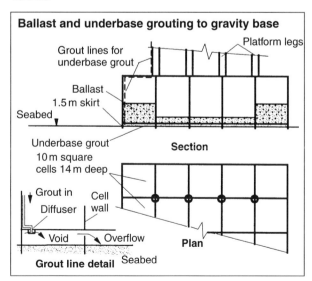

Ballast and underbase grouting to gravity base

Grout lines for underbase grout
Platform legs
Ballast
1.5 m skirt
Seabed
Underbase grout 10 m square cells 14 m deep
Section
Grout in
Diffuser
Cell wall
Void
Overflow
Plan
Grout line detail
Seabed

Monitoring of grout density by mud balance and densimeter recording at the mixer and flow and pressure data recorders at the pump is essential; grout bleed is also regularly checked.

Grout cubes are taken for UCS and shear testing.

10.2 Improving Pile Performance

The performance of bored cast-in-place concrete piles may be improved by grouting either at the base of the pile or along the shaft as part of the structural design.

The principle of base grouting by design is to overcome disturbance of the soil due to drilling, by compacting the soil below the pile and any drilling debris left in the pile. In addition it can reduce settlement of large-diameter piles by increasing base stiffness, leading to shorter piles being required (Yeats and O'Riordan, 1989). However, base grouting may not increase the overall capacity of a pile (Fleming, 1993) and some of the stiffening benefit is probably due to grout seeping up the shaft.

Suitable for reducing settlement in end-bearing piles in sands and gravels and weak rock using compaction grouting at the pile toe. Some improvement in shaft friction is possible by permeation grouting. Fracture grouting of the pile skin has been used, but the technique is not recommended as control is difficult.

Ground investigation will be as for routine pile design. Where remedial actions using grouting are considered, a forensic examination of the structure and foundation elements will be required to determine the causes of damage and a risk analysis must be undertaken for the prospective treatment.

Design for improving pile performance should preferably be based on full-scale site tests. There must be adequate shaft friction available to contain the grout pressure applied at the base and a large enough base area on which the high injection pressure necessary can act.
Compaction grouting is effective in stiffening thick sand layers below the pile. To avoid unwanted delayed consolidation effects, the compacted bulb should extend around the pile toe and vertically up the shaft for a short distance.
Fissure grouting and treatment of voids should be undertaken prior to pile installation.

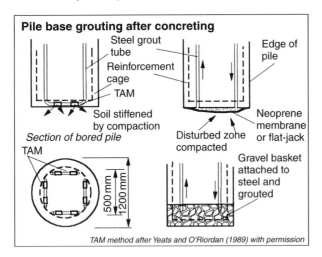

Pile base grouting after concreting

Steel grout tube
Reinforcement cage
TAM
Edge of pile
Soil stiffened by compaction
Section of bored pile
TAM
Neoprene membrane or flat-jack
Disturbed zone compacted
Gravel basket attached to steel and grouted
500 mm
1200 mm

TAM method after Yeats and O'Riordan (1989) with permission

Construction methods for stiffening the pile base post-construction by compaction grouting points built into the pile include:

- grouting using a stiff cement paste through TAMs fixed to the base of the reinforcement cage
- injecting cement grout into flat-jacks installed at the base of the pile
- injecting a gravel basket placed at the base of the pile.

Using the TAM method, packers may be placed at the bottom of the tube either side of the sleeves to limit the volume of grout required or valves may be placed at the top of the injection tube. The tube should be flushed as soon as the pile concrete reaches initial set.

A small volume (0.5–1 m³) of grout paste is injected slowly (1–3 l/min), 14–28 days after concreting, at pressures approximating the base working stress. This can result in pressures several times greater than the overburden resistance and care is needed to avoid fracturing of the ground and uplift of the surface which may adversely affect adjacent piles. Short piles are not amenable to base grouting for this reason. Some consolidation of the base soil can be expected during slow drainage using the pre-placed TAM method.

Monitoring of the pile movements during grouting is essential; the pressure at the base will push the soil down and the pile cap up by different amounts depending on the shaft friction activated. The uplift (say, <1 mm) is the controlling factor.

Other post-construction methods of pile improvement are:

- drilling through the pile and fissure or permeation grouting the soil below and around the toe
- compaction grouting using grout bags inserted into drilled holes below the hardened pile.

PROTECTING AND RENOVATING PILES

Although conventional underpinning remains the main method for foundation renovation, grouting may also be considered as a remedial action for damage caused to piles or as a protective measure against potential damage (van der Stoel, 2001) from:

- increased dragdown on piles in embankments and fills
- lateral displacement alongside excavations
- displacements near tunnels
- earthquakes
- dewatering
- additional superstructure loading.

Remedial work should only be undertaken with full knowledge of the causes of failure and understanding of the effects of the remedial actions. Account must be taken of potential redistribution of pile loads between grouted and un-grouted piles under a structure.

Methods which have been used include the following:
Permeation grouting around and below the pile can improve friction and bearing capacity and reduce the sensitivity of the pile to movement during adjacent construction. Piles may be protected while other works are carried out in the vicinity by permeation grout curtains using TAMs, subject to the appropriate restrictions on soil grading, grout viscosity, injection rates and pressures being observed. Casing is essential when installing TAMs in the vicinity of piles. Injection through a single driven injection point close to the pile toe can improve the stiffness of existing piles.
Jet grouting under piles is likely to cause unacceptable displacements during injection if significant replacement of soil is allowed. Once grout has set, some improvement in capacity in most soils can be achieved. The difference in stiffness of jet grout extensions to end-bearing piles and the original pile will have to be considered, as will the structure stiffness itself, when designing pile renovations and improvements. Jet grout curtains to protect piles are effective if placed >1 m from the piles, but all such injections must be carried out with extreme caution (Sheen, 2001–2).
Compaction grouting can be performed through holes adjacent to piles around the base. Again, caution is required to avoid adverse effects during drilling and grouting.
Crack injection using epoxy resin is not an effective remedial measure for suspect piles.

Rotary percussive track drill forming grout holes for dam cut-off

11 PLANT FOR GEOTECHNICAL PROCESSES

11.1 Drilling

Most of the geotechnical processes considered in this *Introduction* require some form of drilling – either solely for ground investigation and monitoring of the process or for both the investigation and implementation of the process (e.g. groundwater control and grouting).

ROTARY DRILLING

This is the most widely used method of drilling, with a wide range of techniques available for construction, mining, water wells, exploration and geotechnical work. The following notes give some general examples of rigs and indicative performance figures. The rotary action is either by a power-driven rotary 'table' turning a square kelly bar with drill rods and bit down the hole for large-diameter pile holes or more generally by a tophead-drive pneumatic or hydraulic motor turning drill rods and bit at variable speeds up to 500 r.p.m.

Chuck drive rigs, which provide good control of penetration and rotation at high revolutions (up to 700 r.p.m.) for diamond rock coring, cannot produce holes at the rate required for routine geotechnical contracts.

Rotary drilling applications for geotechnical work include:

- diamond core drilling and 'open' hole probing to determine geological structure for ground investigations
- well drilling, grout holes, rock anchors
- pile drilling and pile anchor holes
- augering
- jet grouting.

TOPHEAD-DRIVE ROTARY DRILLING

Large truck-mounted hydraulic tophead-drive drill capabilities for construction	
Operational headroom (under swivel)	10 m
Power swivel torque	12 kNm
Pull-down capacity	75 kN
Hoist load	150 kN
Rotation speed available	10–160 r.p.m.
Maximum hole diameter	300 mm at 700 m depth for this rig, but the range is wide

Drilling tool selection is the key to successful drilling of most formations.

Drill bits for routine rotary drilling of straight holes should be loaded with sufficient weight to ensure penetration while maintaining the drill string in tension as far as possible. The bit rotation speed should be compatible with the weight – the lighter the drill string the higher the r.p.m. This will also extend bit life.

The weight is provided by the drill string and 'drill collars' (heavy walled pipe with an outside diameter less than the hole size) placed just above the bit or by pull-down using the rig hydraulic power. The amount of weight to use will depend on the type of drill bit and formation:

- Hard rock, say, basalt, will require diamond coring, heavily weighted tungsten carbide tricone roller bits or preferably hammer drills and bits to drill open (i.e. unlined or uncased) holes. A general rule for bit loading is 1 tonne per 25 mm of bit diameter for tricone and plate-type bits set with rotating cutters.
- Softer rocks, say, chalk and mudstone, can be drilled with 'drag' or 'fishtail' bits, or 'two-cone' bits requiring less weight. The three-way bit makes a better hole in shattered and partly consolidated formations.

- Clay and sandy overburden can be drilled with continuous flight augers, plate augers and bucket augers.
- Coarse overburden usually has to be cased to prevent hole collapse, but an unlined hole can be drilled in some formations by reverse circulation, using air, water, drilling mud or various flushing combinations.

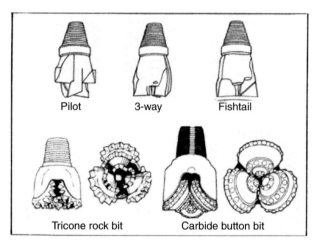

Rotary drilling bits

Pull-down capability is available on most hydraulically powered tophead-drive drills as a means of adding weight to the drill bit. Care must be taken when applying positive feed to the bit in this way to ensure that the rods are not put into excessive compression to cause bending and misalignment of the drill hole.

Most diamond drilling rigs also have positive feed. The driller therefore has to control feed speed to avoid

Rotary hydraulic top-drive rig drilling water supply wells

'crowding' the core bit, together with rotation speed and torque in order to optimise the core recovery.

Drill rods or drill pipe to which the bit is attached make up the drill string and come in a variety of sizes and standards, generally based on API (American Petroleum Institute) nomenclature for rods and connections. (Some British Standard diamond drilling sizes are given in 1.7.) The size must be related to the diameter of the bit so that the annulus between the rod and the hole allows adequate up-hole flushing velocity to clear the drill cuttings from the bottom of the hole and cool the bit. For example, the largest tricone bit recommended on a 2-3/8″ API IF rod is 7-7/8″ (200 mm). It is possible to use larger drag bits compared with drill string diameter in softer formations.

For large-diameter holes in rock and generally where the drill string diameter is small in relation to the hole, it will usually be necessary to use 'stabilisers' and 'reamers' in the drill string to ensure that the rod remains central in the hole.

Flushing media and circulation methods also require careful selection: water, air, mud, foam and combinations with additives are used to meet specific ground conditions and hole diameters. The critical parameters in the field when considering flushing and circulation are:

- ground conditions
- flow rates to provide adequate up-hole injection pressure to indicate changes in the formation being drilled
- restrictions hampering flow, such as reduced diameter for flow through couplings, reduced area between bit face and the ground.

Pile-top rotary drill for anchor pin drilling

Water flush is the most common drilling fluid, pumped through a swivel down the rotating drill string, with the circulation up the annulus to a settlement tank – 'conventional circulation' – to remove relatively coarse cuttings. Typical water flushing requirements for anchors,

mini-piling and core drilling are 100–200 l/min at 20–50 bar with up-hole velocity of 0.5 m/s.

Track-mounted rotary drill forming holes for tie-back anchors

Air flush is mainly used in hard rock and coal measures and possibly in stiff clay where hole collapse does not occur; also where no water supply is available for flushing or mixing drilling mud. Air is not practicable for uncased holes in unconsolidated strata. Air at a suitable pressure (from 6 to 10 bar for deep holes and volume of 15–25 m³/min for normal applications) is circulated as for conventional water flush. Up-hole velocity required for air can be 30 times higher than for water flush, but penetration rates are also much higher.

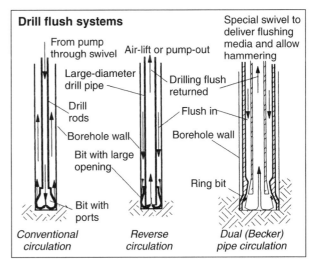

Drill flush systems

Special swivel to deliver flushing media and allow hammering

From pump through swivel · Air-lift or pump-out

Large-diameter drill pipe

Drilling flush returned

Drill rods

Flush in

Borehole wall

Borehole wall

Bit with large opening

Ring bit

Bit with ports

Conventional circulation · *Reverse circulation* · *Dual (Becker) pipe circulation*

Air drilling produces fine drill cuttings, and dust control is essential. The basic method is to cover the top of the hole with a cap and pipe the dust to a cyclone which separates and collects the solids from the return air, which is then discharged to the atmosphere. The cyclone system should not significantly increase air consumption.

When water is encountered in the hole, the hole will become 'sticky', preventing clean removal of cuttings. Removing large quantities of inflowing water with air is costly.

Foaming agents can be injected into the air stream to assist in hole cleaning and suppressing dust. Similarly, the injection of gels such as bentonite and polymers into the air stream can be effective in reducing air loss in porous and fractured formations and where a large volume of water is removed during drilling.

Triple jet grout monitor on drill rods

Foam-injection pumps will deliver 20 l/min at pressures of 2–50 bar to ensure adequate cleaning. Up-hole velocity is less than for water flush, say, 0.2 m/s, and some reduction in penetration rate can be expected.

Mud flushing with bentonite slurry is used as a means of stabilising a drill hole without the use of temporary casing. Density will be 1.04–1.06 tonne/m^3 with a Marsh cone viscosity of 32–38 s (5.2). Up-hole velocity using bentonite mud is 0.4 m/s to remove coarse cuttings.

However, the filter cake which forms on the sides of the hole may impede the effectiveness of the hole as a dewatering well or as a grout injection hole unless considerable development is carried out to remove the mud. Alternatively, a 'biodegradable' mud which reverts to a non-thixotropic fluid within a few days can be used to reduce development and cleaning time.

Mud storage/settlement tanks are required at the surface, usually in conjunction with shakers, de-sanders and centrifuges, to clean the mud before it is returned to the hole – gritty mud increases bit wear.

Drilling with mud is a skilled operation requiring attention to rheology – viscosity, density, gel strength, etc. – including the right selection of constituents and additives for the particular strata.

Reverse circulation methods of rotary drilling provide for the flushing media to be fed down the annulus between the hole wall and the drill pipe or rod and then extracted through the drill bit and up the drill pipe. This is the reverse of 'conventional' flushing down the drill pipe and up the annulus.

The benefits are a more stable hole in soft unconsolidated formations without the use of drilling mud, and large cuttings can be brought up the drill pipe (subject to diameter). This performance is possible because of the high velocity and volume of water flowing up the pipe, particularly when using air-lift principles to remove the flush and cuttings. Holes of 250–1,500 mm diameter can be drilled to 200 m depth using 200 mm diameter flanged drill pipe.

If work is being carried out over water (e.g. for large pile anchor pins) there will be an adequate supply of flush water available (up to 8,000 l/min). Elsewhere large settling pits, to take up to 100 m^3 of water, will be needed so that once the cuttings brought up the drill pipe and discharged into the pit have settled, the cleaned water can be decanted and returned to the hole annulus.

The high volume of water necessary makes the method unsuitable for accurate assessment of the formation being drilled.

Closed-circuit reverse circulation, based on dual-wall drill pipe, has been developed to use both water and air flushing in unstable conditions. It is based on a heavy-duty outer pipe with an inner tube mounted so that there is an annulus for the circulating fluids between the two

tubes. The fluid is pumped down the annulus to the ring drill bit and drill cuttings are returned up the central tube. The outer diameter ranges from 100 mm to over 2,000 mm and can be adapted to most heavy-duty top-drive rotary methods. Advantages are that lost circulation does not affect sample recovery, there is less possibility for hole collapse and artesian heads can be controlled.

Rotary rig drilling wells using reverse circulation

Lost circulation occurs when a stratum is encountered which allows the drilling fluid to penetrate the walls of the borehole instead of returning to the surface with the drill cuttings; this will slow down the drilling progress but more seriously can prevent continued drilling. In severe cases it is necessary to add fibrous material to the drilling fluid to choke off the loss or to grout up the difficult formation and re-drill.

Casing can be successful in curing loss of circulation but the insertion itself requires a large amount of flushing water. It can be time-consuming using only rotary methods in collapsing and coarse overburden formations. A tungsten cutting shoe is fitted to the end of the casing to assist insertion.

Loss of circulation of mud may lead to environmental problems with material reaching water courses.

ROTARY PERCUSSIVE DRILLING

Rotary percussive drilling uses both tophead-drive and down-the-hole hammers (DTHs).

Tophead-drive hammers ('drifters') may be pneumatic (see page 101) or hydraulic and can be mounted on skids, tracks, excavators or trucks with appropriate prime movers. The drill string is in impact compression, producing a tendency to corkscrew unless the drill string is

Rotary percussive track drill on feed beam for soil nails

stabilised when drilling slim holes without casing. Pull-down should be minimised.

Applications include drilling relatively slim holes in rock for wells, probing investigations, grout and anchor holes, mini-piling and rock bolting. Hand-held percussion drills, as used in mining, also have uses in grouting and bolting. Drilling in overburden is not feasible unless casing can be placed before hole collapse or preferably by using one of the proprietary simultaneous drilling and casing methods.

Drill bits are available in a variety of head designs (cross, full-face button and retrac), with differently shaped cutting inserts for different rock conditions with diameters up to 127 mm. Drill strings are 38–50 mm diameter extension rods with special threads to resist the percussion action. Water, air and foam flush can be used in conventional circulation.

Simultaneous drill and case (duplex) equipment uses standard drill string with an under-reaming/pilot bit. The heavy-duty casing tubes and ring bits (outside diameter up to 220 mm) are driven and rotated using a special driving adapter for the tophead hammer.

Typical air-driven tophead-drive rotary percussive drifter capabilities	
Hammer blows	1,950/min
Power swivel torque (up to)	2 kNm
Air consumption (at 7 bar)	5 m³/min
Rotation speed available	5–100 r.p.m.
Maximum open hole diameter	127 mm at 100 m (depending on hoist capacity)

Maximum cased depth for 76 mm casing in uniform soil is 40 m. High-pressure water flush is recommended below the water table. Operational output is 20–25 m/h in coal measures.

A dual-pipe circulation system which can be hammered with or without rotation into overburden to a depth of 50 m is also available (largest standard size 'Becker' pipe: 140 mm outside diameter, 80 mm inside). The torque required is 6 kNm rotating at variable speeds up to 150 r.p.m. and up to 100 blows/min at 10 kNm torque. Air or water flush will produce good samples and the system is useful for installing TAMs in coarse alluvium. Any dual-pipe system is expensive.

Noise control is essential on all percussive rigs. The new 'EU Physical Agents (Noise) Directive' gives a proposed ambient noise limit of 87 dB A and requires workers to wear ear protectors if noise levels exceed 83 dB A. The 'Noise at Work Regulations' (HSE, 1998 and TSO, 2002) currently apply. This means that the tophead hammer

has to be shrouded (muffled), causing some loss of pneumatic power, and that acoustic screens may be needed around the drill.

Hydraulic top-drive rotary percussive rig drilling grout holes for mine infilling

Hydraulically powered tophead-drive power swivels/hammers are now extensively used on the arms of hydraulic excavators for versatile geotechnical drilling operations. Rotation speeds vary up to 250 r.p.m., and with oil pressure at 160 bar, 6 kNm torque can be provided.

Down-the-hole hammer drills are pneumatically operated bottom hole drills that efficiently combine percussive action and rotation. They can be used on any rotary rig – track, truck or skid mounted – with high-pressure air circulation (14 bar) and with the capability of injecting some water to form 'air mist' or foam. The penetration rate depends on the number of blows, air pressure and volume required; the higher, the better the output, typically 35 m/h in limestone for 150 mm holes.

If the power swivel, tracks and feed beam are also powered by air, then an additional 10 m³/min of air at >7 bar will be required.

They operate best in hard rock but cannot cope with large water inflows – when the water pressure equals the air pressure the hammer will cease to work.

Typical DTH drill capabilities	
Hammer blows	400–1,000/min
Power swivel torque needed	0.5–5 kNm
Air consumption	10–25 m³/min
Air pressure	10–20 bar
Rotation speed available	10–50 r.p.m.

Applications are similar to rotary percussive drills. Straight holes are more likely than with tophead hammer rotary percussive rigs as the drill string is not placed under compression. Pull-down should be minimised. Angled holes can be drilled.

Drill bits are similar to full-face rotary percussive bits (usually not cross bits). Hammers are available to suit bits up to 1,100 mm, but air consumption and torque requirements are high (up to 120 m³/min and 50 kNm). It will be essential to provide stabilisers behind the large hammer and possibly 'junk' baskets to collect drill cuttings if adequate up-hole velocity cannot be achieved in a large annulus.

Multiple hammers made up of four separate units within a special housing have also been made for 1,000 mm diameter holes.

Duplex equipment has developed significantly with the advent of large DTH tools. The simultaneous insertion of casing with the DTH is much more efficient than with tophead-drive duplex as the hammer drives the casing shoe without rotation above the under-reaming/pilot bit. Standard equipment will insert 220 mm casing 100 m deep, with conventional or reverse circulation.

The large-diameter hammers have been adapted to duplex drive 1,000 mm diameter casing (wall thickness >12.5 mm) for overburden drilling.

Down-the-hole motors, developed from directional drilling techniques for oil wells, are now available for geotechnical applications such as horizontal well drilling and drilling for service pipes under rivers and roads. The hydraulically powered motor or turbine rotates a drill bit which can be directionally controlled from the surface using a continuous 'survey collar' at the motor.

ROTARY TABLE DRILL

The basic difference between this method of rotation and the tophead-drive rig is that on the rotary table the power unit, either through a mechanical geared connection or through hydraulic motors, drives a horizontal table mounted on a crane or truck. The table has a central key hole which, as it rotates, turns the square or hexagonal kelly bar mated in the key. Below the table, normal drill rods are connected to the kelly and the full range of rotary drill bits and DTH tools can be deployed.

Flushing media can be supplied through swivel connections to the top of the kelly. Inserting casing with rotary table rigs requires space under the table to insert the casing and additional equipment such as a casing oscillator/vibrator with pull-down (crowding) capability.

Applications include reverse circulation, air-lift, rotary and DTH drilling for wells, piling and anchoring.

Mounted on trucks, table rotary drills can drill 1,500 mm holes to 70 m deep; static units with rotary tables up to 1,500 mm can drill 5 m diameter shafts. Crane-mounted units are used for plate auger boring for large-diameter piles.

Hydraulic casing oscillators with pull-down rams or vibrators can be used with most rotary drills – DTH, rotary, hammer grab – to assist in overburden drilling. The oscillator can ensure that the casing (with carbide cutting shoe) is kept close to the bottom of the hole. Diameters up to 2,500 mm are available, with torque of 2,100 kNm at 300 bar operating pressure. They are used for well drilling and piling in difficult formations (e.g. karst).

Typical truck-mounted rotary table rig capabilities	
Operational headroom (under swivel)	12 m
Rotary table diameter	800 mm
Hoist capacity	16 kN
Rotation speed available	15–90 r.p.m.
Hole diameter	700 mm to 400 m deep for this rig but the range is wide – up to 1,500 mm diameter

CONTINUOUS FLIGHT AUGER

CFA rotary drilling (see page 63) is versatile and is used extensively for ground investigation and geotechnical processes in sands, gravels and weak rock, drilling dry

and usually avoiding the need to case holes – except in high water tables. The flight diameters of hollow-stem augers range from 150 mm for mini-piles to 1,200 mm for large CFA piles. High-torque (>200 kNm) tophead-drive power swivels are mounted on torsion-resisting masts up to 25 m, to allow the hole to be drilled in one run.

Noise and vibration levels are low.

Drill bits available include various types of 'V' bit, finger bits and rock augers for soft rock.

CABLE TOOL PERCUSSIVE DRILLING

Here the hole is formed by the impact and cutting action of a long chisel-edged drill bit to break up the strata into small fragments. The vertical reciprocating motion (no powered rotation) of the tool mixes the loosened material into a sludge that is removed by a bailer (similar to the shell and auger method of boring ground investigation holes in 1.7). Water may have to be added.

Casing of up to 1,800 mm diameter can be driven, and 500 mm diameter wells can be bored to 500 m.

Applications are mainly for water supply wells; rigs are relatively slow for

Cable percussive rig drilling dewatering well

the limited diameters and depths required for dewatering (350 mm and <60 m).

Safety precautions
Drilling operations can be hazardous. The BDA (2002) safety manual and training recommendations should be strictly observed.
Operatives should always wear ear-protectors and full safety clothing, safety boots and helmets

A 2 tonne light cable tool (shell and auger) rig is versatile and can quickly install 200 mm well screen in sands and gravels to 20 m.

Mounting can be on a truck, trailer or skids, with masts up to 18 m and 20 tonne hoist capacity.

COSTS

Costs depend on conditions, the formation being drilled, the method of operating and many other factors. In sound, hard rock under the same conditions the DTH system is cheaper than

Rotating cutters for diaphragm wall trench

both the tophead-drive drifter and rotary drilling; auger drilling is cheapest for cohesive soils; duplex drilling is economic in unconsolidated soils.

Diaphragm wall grab

11.2 Grout Mixing

Mixing of grout can be divided into making suspensions of solids and stirring chemical solutions.

Weigh-batching of constituents is generally specified for most grouting applications – volume-batching of water and chemicals of known concentration is normal. All batching units should be independently certified and throughputs recorded either manually or automatically.

Pre-weighed bags or onsite scales can be used for batching small quantities of dry cement, sand, PFA and bentonite. For larger outputs, weigh hoppers with screw feed into the mixer are used; coarse material may be weighed on weighing conveyors to a reasonable accuracy.

Pressurised silos (with compressed air at 1–2 bar) feeding rotating measuring valves can accurately deliver dry fine materials continuously to the mixer unit. This method provides good control of dust around the plant.

Stockpiles of sand and PFA can be batched by weighing the mixer as the constituents are added, but because the materials will be dampened down to control dust it is necessary to take account of moisture contents.

Chemical grout constituents are used in small quantities through volumetric dispensers discharging into stirrers. The dispensers must be accurate to maintain gel time control and constructed so that cross-contamination cannot occur or fluids escape. Proportioning pumps which deliver chemicals to a mixing head at the injection point are available but difficult to control consistently.

Paddle mixers (or pan mixers) are the simplest form and may be used for batch mixing small or large volumes (>2 m^3) of high water:solids ratio grouts of cement and PFA. Paddles can be mounted vertically or horizontally, depending on the way the solids are added to the water; where the open mixing tank is fed with PFA from a loader bucket, horizontal mounting is more efficient.

For vertical shaft mixers, baffles are fitted to the side of the tank to improve mixing. The batch volume used in these bulk mixers should be half the capacity of the mixer tub to ensure good mixing in a reasonable time.

Direct drive by large diesel engine is used for the large units, but pneumatic, electric and hydraulic units are made in various capacities.

Agitation tanks are also paddle mixers, usually with a vertical shaft paddle and side baffles, which are used to hold the grout in homogeneous suspension prior to pumping. Also acts as the header tank for the pump suction. The grout from the bulk mixer is passed through a strainer to prevent lumps of material entering the tank. Where considerable amounts of air may be entrained in the mix, a perforated agitation paddle helps to remove the air. The capacity of agitation tanks should be twice that of the bulk mixer.

Drive may be by pneumatic, hydraulic or electric power at a speed of 50–100 r.p.m.

Stirrers for chemicals are small, rapid paddle mixers (50–200 r.p.m.), with batch size taking account of the gel time, viscosity of the mix components (elastomers, polyurethane, epoxy resins) and the usually slow injection rate. Quick-gelling grouts should be mixed close to the injection point. Expert advice should be obtained from mixer manufacturers.

> **Caution**: when storing, handling and mixing chemicals on site COSSH Regulations must be strictly observed and appropriate protective clothing worn.

Colloidal mixers are required for mixing particulate grout suspensions for injection into soil and rock. The intention is to coat all the particles with the optimum amount of water to ensure that they remain in suspension during pumping, that the chemical reaction will be uniform and that the rheological properties are consistent. This is achieved by the rapid shearing of slurry particles by a rotating impeller and/or rapidly swirling the mix in a vortex around the side of a cylindrical mixing tank. The grout flow properties are considerably improved compared with paddle mixing.

The mixing impeller type can cope with most grout solids, cement, PFA, bentonite and medium-size sand. (Note that the lowest water:cement ratio to achieve chemical hydration is 0.32.) The mixer consists of a cylindrical tank with a mixer housing below the base containing the high-speed disc impeller (1,500 r.p.m.). Batched water is fed into the tank, the impeller started and the constituents added. These are first drawn into the impeller then discharged tangentially into the top of the mixing tank, causing a swirling/shearing action before returning to the impeller unit. Small batches, hand fed, can be mixed in under 1 min, the mix being discharged into the agitation tank. The process can be adapted for larger weigh-batching operation.

Double-drum mixers are useful when mixing strong sand:cement grout (1–3:1). The sand is mixed separately in one drum with a star-shaped impeller in the mixer housing, while the cement is mixed in the second drum with the disc impeller. The cement slurry is then transferred to the sand mix for final mixing before being discharged to the agitation tank for pumping.

Batch volume ranges from 0.1 to 0.5 m^3 with one batch every 1–2 min using 40 kW powered units.

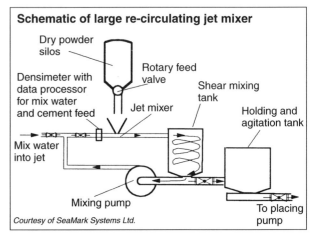

Schematic of large re-circulating jet mixer

Courtesy of SeaMark Systems Ltd.

Jet mixers are more convenient for large continuous outputs of grout from automatic weigh-batchers using several constituents. Here the initial mixing unit is a large venturi jet on to which the dry constituents are fed from rotary feed valves. The slurry is then delivered tangentially into the top of the mixing tank and bottom-discharged to a slurry pump (250 mm self-priming centrifugal unit) which further circulates and shears the mix around the tank 3–4 times before automatically discharging at the pre-set density to the agitation tank.

Outputs of up to 0.6 tonne of grout per minute at density of up to 2 tonne/m^3 are achievable. Simpler jet mixers are used for making bentonite slurry for drilling and pre-mixing for use as a grout additive. A feed hopper provides bentonite powder from bags on to a high-pressure venturi jet which draws in and mixes the slurry before it is discharged to agitation tanks or large re-circulating systems for the required conditioning period.

Pug mills, which will form a soft paste with water, are required for pre-treatment mixing of very fine powders – such as microfine cement.

11.3 Grout Pumping

Grout pumps should provide good control of pressure and flow rates. Positive displacement pumps with long-stroke double-acting pistons are ideal for permeation, but helical screw pumps are generally more common. It is important to avoid pumps which produce uneven pressure peaks. Large centrifugal pumps are suitable for mixing and placing high-volume coarse particle grouts. Output needed will vary from 0.1 to 50 l/s and from 1 to 50 bar pressure, depending on application, grout and depth.

Selection of a pump will be based on:

- grouting method and grouts to be used
- variations in pressure and output required
- control systems at the pump and injection point
- wear resistance needed
- valves and grout pipes compatible with required pressures.

Centrifugal pumps designed to handle coarse slurries are mainly used as mixing pumps but can have applications for bulk infilling of large voids. A typical 3/2 pump for mixing and delivery has an output of 20 l/s at 5 bar. They are not generally suitable for fine control of delivery volume and pressure.

Positive displacement pumps which deliver a set volume of fluid for every stroke of the piston or rotation of the pump screw are favoured for most grout injection work.

Helical rotor (screw) pumps have a stainless steel helical rotor which rotates in a rubber stator with the same helical shape so that 'slugs' of fluid are forced along the stator. Flow rates of 0.1–30 l/s and pressures up to 24 bar are available. The pump will cope with a variety of particulate grouts, including high-viscosity compaction grout and resins – special stators may be needed.

For viscous grouts an auger feed into the rotor ensures even pumping.

Output range of helical rotor pumps
11 kW drive motor at 720 r.p.m. and 50 mm delivery pipe: 0.5–1 l/s at 0–24 bar total head
30 kW drive motor at 540 r.p.m. and 100 mm delivery pipe: 27 l/s at 0–6 bar total head

Piston (or plunger) pumps are available in a wide range of outputs and pressure depending on viscosity and application. Most pumps of this type will handle coarse and abrasive particles (<2 mm) at low pressure provided ball valves are used, but separation of the mix is possible at high pressure. Most chemical grouts can be handled.

Long-stroke pumps at slow speed will provide best control over flow rate and pressure to avoid 'peaking' during permeation. But with modern hydraulic powered units the smaller, shorter-stroke piston pumps are acceptable for most applications. Typical discharge per stroke is 0.5–2 l at around 40 strokes/min.

Double-piston concrete pumps are used for heavy-duty applications with stiff mortars such as compaction grouting, shotcreting (wet process) and pressure pointing. Outputs in such cases are 0.25–1.3 l/s at pressures up to 60 bar with 7.5 kW drive. Delivery distance is 500 m horizontally with fluid grout.

Triplex pumps are used for pumping water to jet mixers and should not be used to handle cement or chemical grouts.

Small (4 kW) pneumatic drive (6 bar) piston pump	
Output range	1 l/s at 70 bar to 0.1 l/s at 150 bar 50 mm plunger; 25 mm delivery

Electro-hydraulic drive (7.5 kW) single-cylinder double-acting piston pump	
Output range	0–3.5 l/s at 2–55 bar 150 mm grout cylinder; 25 mm delivery

Large (25 kW) duplex piston pump	
Output range	63–127 mm liner: 6 l/s at 9 bar to 1 l/s at 35 bar 250 mm stroke; 50 mm delivery

Peristaltic 'flow inducers' are useful for accurate metering of epoxy resins and other difficult fluids. The pump is formed by an elastic neoprene tube clipped to a curved track, where the centre of the curve is concentric with a rotor carrying three rollers. As the rotor turns, the rollers pass over the tube, flattening it against the track at the point of contact. This 'flat' moves around the track, driving the fluid in the tube before it and at the same time creating a suction behind the roller to draw in more fluid. Output depends on tube diameter and rotor speed.

ANCILLARY INJECTION EQUIPMENT

Packers are sealing plugs used for water loss tests during investigations and after grouting, for the injection of grout into rock and with TAM methods. Single packers are placed at the top of a hole in a prepared concrete base for shallow treatment or down the hole to isolate a stratum. Double packers are placed down the hole and seal the test formation or grouting zone at a specific level.

Packers are sealed in the hole either by compression or inflation:

- mechanical outward squeezing of a rubber sleeve
- inflating a rubber bag around the grout tube using a separate air or hydraulic line
- tightening in a smooth intact hole under the applied water pressure.

Sealing packers in broken rock, packer rupture and water escaping around the packer all affect the test results and grouting efficiency.

Data recorders, either integral with the equipment or standalone, should provide real-time information on flow, pressure, volume and time for most grouting work.

Flow meters for batching water can be the basic vane type. For measuring grout flow, non-mechanical types such as electromagnetic flow meters avoid the otherwise inevitable blockages.

Multiple injection points (up to three) from one pump require careful tuning of control ball valves to ensure grout flow, and data recorders should be provided on each injection line.

In-line densimeters are useful in the mixing cycle for high-strength grout; the weighing type is susceptible to movements and the nuclear attenuation type requires monitoring by registered personnel.

The standard mud balance is generally acceptable for most applications as part of the rheology testing regime.

Uplift gauges set either at the surface or at a particular stratum are necessary when ground movement is the limiting factor for injection.

GLOSSARY

As the subject of geotechnics is so wide-ranging, a glossary of all the terms used in geotechnical processes would overwhelm the short text. The following standard glossaries in current use are therefore recommended.

BS 3618: Section 5, 1971 (1987) *Glossary of Mining Terms: Geology*. British Standards Institute, London.

BS 3618: Section 6, 1972 (1987) *Glossary of Mining Terms: Drilling and Blasting*. British Standards Institute, London.

BS 3618: Section 11, 1967 (1987) *Glossary of Mining Terms: Strata Control*. British Standards Institute, London.

BS 6100: Part 2, Section 2.2, *Glossary of Building and Civil Engineering Terms: Sub-section 2.2.1: 1992 (1998) Earthworks, Sub-section 2.2.2: 1990 (1997) Substructures and foundations, Sub-section 2.2.3 1990 (1997) Tunnels*. British Standards Institute, London.

BS EN 12715: 2000 *Execution of Special Geotechnical Works – Grouting*. British Standards Institute, London. [Contains a glossary of grouting terms.]

Scott J (1991) *Dictionary of Civil Engineering*, 4th edn. Penguin Books, London.

W S Atkins Consultants Limited (1997) *Glossary of Terms and Definitions Used in Grouting: Proposed Definitions and Preferred Usage*, CIRIA Report PR61. Construction Industry Research and Information Association, London.

Somerville S H and Paul M A (1983) *Dictionary of Geotechnics*. Butterworth, London.

SELECTED BIBLIOGRAPHY

Contracts

O'Reilly M (1999) *Civil Engineering Construction Contracts*. Thomas Telford, London.

Dewatering

Cashman P M and Preene M (2001) *Groundwater Lowering in Construction – A Practical Guide*. Spon Press, London.

Engineering geology

Blyth F G H and de Freitas M H (1984) *A Geology for Engineers*, 7th edn. Butterworth-Heinemann, Oxford.

Environmental geotechnics

Hiller D M and Crabb G I (2000) *Groundborne Vibration Caused by Mechanised Construction Works*. Thomas Telford, London.

Sarsby R W (2000) *Environmental Geotechnics*. Thomas Telford, London.

Environmental law

Stubbs A and Derring C (2002) *Environmental Law for the Construction Industry*. Thomas Telford, London.

Foundations generally

Tomlinson M J (2001) *Foundation Design and Construction*, 7th edn. Prentice-Hall, London.

Ground improvement

Moseley M P (ed.) (2004) *Ground Improvement*, 2nd edn. Spon Press, London.

Grouting

Damone P L J and Jefferis S A (eds) (1995) *Structural Grouts*. Blackie Academic & Professional, Glasgow.

Hellawell E E, Rawlings C G and Kilkenny W M (1997) *Geotechnical Grouting: A Bibliography*, CIRIA Project Report PR60. Construction Industry Research and Information Association, London.

Lime stabilisation

Rogers C D F, Glendinning S and Dixon N (eds) (2001) *Lime Stabilisation*. Thomas Telford, London.

Rock stabilisation

Wyllie D C and Mah C W (2004) *Rock Slope Engineering*, 4th edn. Spon Press (up-dated version of Hoek and Bray (1981) (q.v)).

Safety

Health and Safety Executive (2001) *Managing Health and Safety in Construction*. HSE Books, London.

Statistics

Baecher G B and Christian J T (2003) *Reliability and Statistics in Geotechnical Engineering*. John Wiley & Son Inc., New York.

Tunnelling

Burland J B, Standing J R and Jardine F M (eds) (2001) *Building Response to Tunnelling Case Studies from Construction of the Jubilee Line Extension*, CIRIA Special Publication 200. Thomas Telford, London.

REFERENCES

Works by the following bodies are listed under the abbreviated form.

AGS   Association of Geotechnical and Geoenvironmental Specialists (UK)
API   American Petroleum Institute
ASCE   American Society of Civil Engineers
BDA   British Drilling Association
BRE   Building Research Establishment (UK)
BS   British Standards (Institute)
CIRIA   Construction Industry Research and Information Association (UK)
DETR   Department of the Environment, Transport and the Regions (UK)
HMSO   Her Majesty's Stationery Office
HSE   Health and Safety Executive (UK)
ICE   Institution of Civil Engineers (UK)
OCMA   Oil Companies Materials Association (UK)
RDGC   Recherche Développement en Génie Civil (France)
TSO   The Stationery Office

AGS (1998) *AGS Guide: The Selection of Geotechnical Soil Laboratory Testing*. Association of Geotechnical and Geoenvironmental Specialists, Beckenham, Kent, UK.

AGS (2000) *Electronic Transfer of Geotechnical Data from Geotechnical Investigations*, 3rd edn. Association of Geotechnical and Geoenvironmental Specialists, Beckenham, Kent, UK.

AGS (2002) *Safety Manual for Investigation Sites*. Association of Geotechnical and Geoenvironmental Specialists, Beckenham, Kent, UK.

Al-Tabbaa A, Evans C W and Wallace C J (1998a) 'Pilot *in situ* auger mixing treatment of a contaminated site. Parts 1: Treatability study'. *Geotechnical Engineering* 131(1): 52–9. Institution of Civil Engineers, London.

Al-Tabbaa A, Evans C W and Wallace C J (1998b) 'Pilot *in situ* auger mixing treatment of a contaminated site. Part 2: Site trial'. *Geotechnical Engineering* 131(2): 89–95. Institution of Civil Engineers, London.

Al-Tabbaa A, Evans C W and Wallace C J (2000) 'Pilot *in situ* auger mixing treatment of a contaminated site. Part 3: Time-related performance'. *Geotechnical Engineering* 143(2): 103–14. Institution of Civil Engineers, London.

Al-Tabbaa A, Evans C W and Wallace C J (2003) 'Pilot *in situ* auger mixing treatment of a contaminated site. Part 4: Performance at five years'. *Geotechnical Engineering* 155(3): 187–202. Institution of Civil Engineers, London.

API (1990) *Recommended Practice RP13B-1: Standard Procedure for Testing Drilling Fluids*. American Petroleum Institute, Washington DC 20005.

Auld F A and Harris J S (1995) 'Engineering design of frozen ground works' in J S Harris (ed.) *Ground Freezing in Practice*. Thomas Telford, London, pp. 70–86.

Baker A C J and James A N (1990) 'Three Valleys Water Committee: tunnel connection to Thames Water reservoirs' in *Proceedings of Institution of Civil Engineers*, Part 1. London, pp. 929–54.

Barley A D (1997) 'The single bore multiple anchor system' in G S Littlejohn (ed.) *Ground Anchorages and Anchored Structures*. Thomas Telford, London, pp. 65–75.

Barley A D and McBarron P L (1997) 'Field trials on four high capacity removable multiple anchors founded in marine sand fill and in completely decomposed granite' in G S Littlejohn (ed.) *Ground Anchorages and Anchored Structures*. Thomas Telford, London, pp. 148–57.

Bauman V and Bauer G E A (1974) 'The performance of foundations on various soils stabilised by the vibro-compaction method'. *Canadian Geotechnical Journal* 11: 509–30. National Research Council of Canada.

BDA (2002) *Health and Safety Manual for Land Drilling – Code of Safe Drilling Practice*. British Drilling Association, UK.

Bell A L (1982) 'A cut off in rock and alluvium at Asprokremmos dam' in *Proceedings of Conference on Grouting in Geotechnical Engineering, New Orleans*. American Society of Civil Engineers, New York, pp. 172–86.

Bell A L (1993) 'Jet grouting' in M P Moseley (ed.) *Ground Improvement*. Blackie Academic & Professional, Glasgow, pp. 148–74.

Bell F G and Culshaw M G (2001) 'Problem soils: a review from a British perspective' in I Jefferson, E J Murray, E Faragher and P R Fleming (eds) *Problematic Soils*. Thomas Telford, London, pp. 2–35.

Beresford J J, Cashman P M and Hollamby R G (1989) 'Merits of polymeric fluids as support slurries' in *Proceedings of Conference on Piling and Deep Foundations*. A A Balkema, Rotterdam, Vol. 1, pp. 3–11.

Bergado D T, Balasubramaniam A S, Pakawaran M A P and Kwunpreuk W (2000) 'Electro-osmotic consolidation of soft Bangkok clay with prefabricated vertical drains'. *Ground Improvement* 104(4): 53–63. Thomas Telford, London.

Beveridge J P and Rankin W J (1995) 'Role of engineering geology in NATM construction' in M Eddlestone, S Walthall, J C Cripps and M G Culshaw (eds) *Engineering Geology of Construction*, Engineering Geology Special Publication No. 10. Geological Society of London, pp. 255–68.

Billington C J and Tebbett I E (1980) 'The basis for a new design formula for grouted jacket to pile connections' in *Proceedings of the Offshore Technology Conference*. Houston, TX, paper OTC3788, pp. 449–58.

Binns A (1998) 'Rotary coring in soils and soft rocks for geotechnical engineering'. *Geotechnical Engineering* 131(2): 63–74. Institution of Civil Engineers, London.

Bjerrum L and Eide O (1956) 'Stability of strutted excavations in clay'. *Geotechnique* 6(1): 32–47. Institution of Civil Engineers, London.

Blyth F G H and de Freitas M H (1984) *A Geology for Engineers*, 7th edn. Butterworth Heinemann, Oxford.

Bowles J E (1988) *Foundation Analysis and Design*, 4th edn. McGraw-Hill International Editions, New York.

BRE 458 (2003) *Specifying Dynamic Compaction*. Building Research Establishment, Watford, UK.

BRE Digest 240 (1993) *Low-Rise Buildings on Shrinkable Clay Soils: Part 1*. Building Research Establishment, Watford, UK.

BRE Digest 348 (1989) *Site Investigation for Low-Rise Building: The Walk-over Survey*. Building Research Establishment, Watford, UK.

BRE Digest 412 (1996) *Desiccation in Clay Soils*. Building Research Establishment, Watford, UK.

Brice G J and Woodward J C (1984) 'Arab Potash solar evaporation system: design and development of a novel membrane cut-off wall' in *Proceedings of Institution of Civil Engineers*, Part 1. London, pp.185–205.

British Cryogenics Council (1991) *Cryogenics Safety Manual*, 3rd edn. Butterworth, Oxford.

Bromhead E N (1992) *The Stability of Slopes*, 2nd edn. Chapman and Hall, London.

Broms B (1993) 'Lime stabilisation' in M P Moseley (ed.) *Ground Improvement*. Blackie Academic & Professional, Glasgow, pp. 65–99.

Broms B (1994) *Stabilisation of Soil with Lime Columns*, Design Handbook. Royal Institution of Technology, Oslo.

Brown E T (ed.) (1981) *Rock Characterisation, Testing and Monitoring. ISRM Suggested Methods*. Pergamon Press, London.

Bruce D A (1997) 'The stabilisation of concrete dams by post-tensioned rock anchors – the state of American practice' in G S Littlejohn (ed.) *Ground Anchorages and Anchored Structures*. Thomas Telford, London, pp. 508–21.

Bruce D A, Hague S T and Hitt R (2001a) 'Treatment by jet grouting of a bridge foundation on karstic limestone' in T L Brandon (ed.) *Foundations and Ground Improvement*, Geotechnical Special Publication 113. American Society of Civil Engineers, New York, pp. 145–59.

Bruce D A, Traylor R P and Lolcama J (2001b) 'Sealing of a massive water flow through karstic limestone' in T L Brandon (ed.) *Foundations and Ground Improvement*, Geotechnical Special Publication 113. American Society of Civil Engineers, New York, pp. 160–74.

BS 1377: 1990 *Methods of Test for Soils for Civil Engineering Purposes*. British Standards Institute, London.

BS 1924: 1990 *Stabilised Materials for Civil Engineering Purposes*. Part 1: *General Requirements, Sampling, Sample Preparation and Tests on Materials before Stabilisation*. Part 2: *Methods of Test for Cement-Stabilised and Lime-Stabilised Materials*. British Standards Institute, London.

BS 3892: 1997 Part 3: *Specification for Pulverised-Fuel Ash for Use in Cementitious Grouts*. British Standards Institute, London.

BS 4019: 1993 Parts 3–4. *Rotary Core Drilling Equipment – Specification for System A Metric Units & Specification for System A Inch Units*. British Standards Institute, London.

BS 4486: 1980 *Specification for Hot Rolled and Hot Rolled and Processed High Tensile Alloy Steel Bars for the Prestressing of Concrete*. British Standards Institute, London.

BS 5228: 1997 Parts 1–4: *Noise and Vibration Control on Construction and Open Sites*. British Standards Institute, London.

BS 5607: 1988 *Safe Use of Explosives in the Construction Industry*. British Standards Institute, London.

BS 5896: 1980 *Specification for High Tensile Steel Wire and Strand for the Prestressing of Concrete*. British Standards Institute, London.

BS 5930: 1999 *Code of Practice for Site Investigations*. British Standards Institute, London.

BS 6031: 1981 *Code of Practice for Earthworks*. British Standards Institute, London.

BS 6472: 1992 *Guide to Evaluation of Human Exposure to Vibration in Buildings (1 Hz to 80 Hz)*. British Standards Institute, London.

BS 6906: 1995 Parts 2, 3, 5 and 7: *Methods of Test for Geotextiles* (also BS EN ISO 10319: 1996, BS EN 918: 1996, and BS EN 12236: 1996 replacing Parts 1, 4 and 6 respectively). British Standards Institute, London.

BS 7022: 1989 *Guide for Geophysical Logging of Boreholes for Hydrogeological Purposes*. British Standards Institute, London.

BS 8002: 1994 *Code of Practice for Earth Retaining Structures*. British Standards Institute, London.

BS 8004: 1986 *Code of Practice for Foundations*. British Standards Institute, London.

BS 8006: 1995 *Code of Practice for Strengthened/Reinforced Soils and Other Fills*. British Standards Institute, London.

BS 8008: 1996 *Safety Precautions and Procedures for the Construction and Descent of Machine-Bored Shafts for Piling and Other Purposes*. British Standards Institute, London.

BS 8081: 1989 *Code of Practice for Ground Anchorages* (partially superseded by BS EN 1537: 2000). British Standards Institute, London.

BS 10175: 2001 *Investigation of Potentially Contaminated Sites – Code of Practice*. British Standards Institute, London.

BS EN 1537: 2000 *Execution of Special Geotechnical Works – Ground Anchors*. British Standards Institute, London.

BS EN 1538: 2000 *Execution of Special Geotechnical Works – Diaphragm Walls*. British Standards Institute, London.

BS EN 12715: 2000 *Execution of Special Geotechnical Works – Grouting*. British Standards Institute, London.

BS EN 12716: 2001 *Execution of Special Geotechnical Works – Jet Grouting*. British Standards Institute, London.

prEN 13968: 2002 (draft for comment) *Geosynthetic Barriers – Characteristics Required for Use in the Construction of Solid Waste Storage and Disposal Sites and Storages for Hazardous Solid Materials*. British Standards Institute, London.

Burford D and Charles J A (1992) 'Long-term performance of houses built on opencast mining backfill at Corby' in *Proceedings of Conference on Settlement of Structures*. Pentech Press, Cambridge, pp. 54–67.

Burland J B, Standing J R and Jardine F M (eds) (2001) *Building Response to Tunnelling. Case Studies from the Jubilee Line Extension, London,* CIRIA Special Publication 200. Thomas Telford, London.

Card G B and Carter G R (1995) 'Case history of a piled embankment in London's Docklands' in M Eddlestone, S Walthall, J C Cripps and M G Culshaw (eds) *Engineering Geology of Construction*, Engineering Geology Special Publication No. 10. Geological Society of London, pp. 79–84.

Caron C (1963) 'The development of grouts for the injection of fine sands' in *Symposium on Grouts and Drilling Muds in Engineering Practice*. Butterworths, London. pp. 136–41.

Casagrande L (1952) 'Electro-osmosis stabilisation of soils'. *Journal of the Boston Society of Civil Engineers* 39: 51–83.

Cashman P M and Preene M (2001) *Groundwater Lowering in Construction. A Practical Guide*. Spon, London.

Cedergren H R (1989) *Seepage, Drainage and Flow Nets*, 3rd edn. Wiley, New York.

Chandler R J and Forster A (2001) *Engineering in Mercia Mudstone*, CIRIA Report C570. Construction Industry Research and Information Association, London.

Chapman P J, Wrightam V H, Ince G and Lewis R P (1992) 'Channel Tunnel. Part 1: Tunnels; Machine-driven tunnels' in *Proceedings of Institution of Civil Engineers*, London, pp. 55–86.

Charles J A and Watts K S (2002) *Treated Ground – Engineering Properties and Performance*, CIRIA Report C572. Construction Industry Research and Information Association, London.

Chisholm F and Kohen S (1954) 'The measurement of plastic flow properties of drilling mud'. *The Petroleum Engineer* (Apr.): 425–9. Houston Texas.

Chummar A V (1998) 'Problems in ground improvement by stone column' in *Second International Conference on Ground Improvement Techniques*. CI Premier PET Ltd, Singapore, pp. 513–21.

Clarke B G (1996) 'A practical guide to the derivation of undrained shear strength from pressuremeter tests' in C Craig (ed) *Advances in Site Investigation Practice*. Thomas Telford, London (ed.) pp. 559–70.

Clarke B G (1996–7) 'Pressuremeter testing in ground investigation', Parts 1 and 2. *Geotechnical Engineering* (Apr. 1996) 119: 96–108 and (Jan. 1997) 125: 42–52. Institution of Civil Engineers, London.

Clayton C R I (1993) *The Standard Penetration Test (SPT): Methods and Use*, CIRIA Report R143, Construction Industry Research and Information Association, London.

Clayton C R I (2001) *Managing Geotechnical Risk*. Institution of Civil Engineers, London.

Clayton C R I, Matthews M C and Simons N E (1995) *Site Investigation*, 2nd edn. Blackwell Science, London.

Clayton C R I, Hope V S, Heyman G, van der Berg J P and Bica A (2000) 'Instrumentation for sprayed concrete lined soft ground tunnels'. *Geotechnical Engineering* 143(3): 119–30. Institution of Civil Engineers, London.

Cole K W (1993) 'Building over abandoned shallow mines. Paper 1: Considerations of risk and reliability'. *Ground Engineering,* 26(1): 34–6. Thomas Telford, London.

Cole K W and Statham I (1992a) 'General (areal) subsidence above partial extraction mines, Part 1'. *Ground Engineering,* 25(2): 45–55. Thomas Telford, London.

Cole K W and Statham I (1992b) 'General (areal) subsidence above partial extraction mines, Part 2'. *Ground Engineering,* 25(3): 36–40. Thomas Telford, London.

Cole R G, Carter I C and Schofield R J (1994) 'Staged construction at Benutan Dam assisted by vacuum eductor wells' in *Eighteenth International Conference on Large Dams, Durban, South Africa*, Vol. III pp. 625–40. International Commission on Large Dans Paris.

Collins S P and Deacon W G (1972) 'Shaft sinking by ground freezing: Ely Ouse-Essex scheme' in *Supplement to Proceedings of the Institution of Civil Engineers*. London, paper 7506S, pp. 367–78.

de Paoli B, Tornaghi R and Bruce D A (1989) 'Jet grout stabilisation of peaty soils under a railway embankment in Italy' in F H Kulhawy (ed.) *Foundation Engineering, Current Principles and Practice*, American Society of Civil Engineers Conference, Evanston, IL. American Society of Civil Engineers, New York, pp. 272–90.

Domone P L J (1990) 'The properties of low strength silicate/portland cement grouts'. *Cement and Concrete Research* 20(1): 25–35. Pergamon Press UK.

Domone P L J (1994) 'Hardened properties of Portland cement grouts' and 'Other types of grout' in P L J Domone and S A Jefferis (eds) *Structural Grouts*. Blackie Academic & Professional, Glasgow, pp. 58–93, 94–108.

Douglas T H and Arthur L J (1983) *A Guide to the Use of Rock Reinforcement in Underground Excavations*, CIRIA Report 101. Construction Industry Research and Information Association, London.

Driscoll F G (1986) *Ground Water and Wells*, 3rd edn. Johnson Division, UOP Inc., Saint Paul, MN.

EA (1999) *Technical Support Materials for the Regulation of Radioactively Contaminated Land*, Research and Development Report P307. Environment Agency, The Stationery Office, London.

Eglinton M S (1987) *Concrete and its Chemical Behaviour*. Thomas Telford, London.

Eurocode 7 Part 1 DD ENV 1997–1: 1995 *Geotechnical Design – General Rules*. British Standards Institute, London.

Eurocode 7 Part 2 (draft): (2000) *Geotechnical Design Assisted by Laboratory Testing*. British Standards Institute, London.

Eurocode 7 Part 3 (draft): (2000) *Geotechnical Design Assisted by Field Testing*. British Standards Institute, London.

Evans G V, Parsons T V and Wallace M R G (1980) 'Nuclear grout monitoring on offshore platforms' in *Proceedings Offshore Technology Conference*. Houston, TX. Paper OTC 3791, pp. 477–84.

Ewert F-K (1994) 'Evaluation and interpretation of water pressure tests' in A L Bell (ed.) *Grouting in the Ground*. Thomas Telford, London, pp. 141–61.

Ewert F-K (1996) 'Considerations on grouting of Karstic limestone at dam sites'. *Dam Engineering* 7(1): 3–33. Read Business Publications, UK.

Farmer I W (1975) 'Electro-osmosis and electro-chemical stabilisation' in F G Bell (ed.) *Methods of Treatment of Unstable Ground*. Newnes-Butterworth, London, pp. 26–36.

Flemming W G K (1993) 'The improvement of pile performance by base grouting' in *Proceedings of the Institution of Civil Engineers*, 97. London, pp. 88–93.

Fookes P G (1997) 'Geology for engineers: the geological model, prediction and performance'. *Quarterly Journal of Engineering Geology* 30(4): 293–424. Geological Society of London.

Fookes P G, Baynes F and Hutchinson J (2001) 'Total geological history: a model approach to understanding site conditions'. *Ground Engineering* 34(3): 42–7. Emap Construct, London.

Francescon M and Twine D (1994) 'Treatment of solution features in Upper Chalk by compaction grouting' in A L Bell (ed.) *Grouting in the Ground*. Thomas Telford, London, pp. 327–47.

Freeze R A and Cherry J A (1979) *Groundwater*. Prentice-Hall Inc., NJ.

Gaba A R, Simpson B, Parry W and Beadman D R (2003) *Embedded Retaining Walls – Best Practice Guide*, CIRIA Report C580. Construction Industry Research and Information Association, London.

Greenwood D and Thomson G H (1984) *Ground stabilisation: Deep Compaction and Grouting*. Thomas Telford, London.

Griffith J S (ed.) (2001) *Land Surface Evaluation for Engineering Purposes*, Engineering Geology Special Publication 18. Geological Society of London.

HA 68/94 (1994) Advice Note *Design Methods for the Reinforcement of Highway Slopes by Reinforced Soil and Soil Nailing Techniques*. Highways Agency, The Stationery Office, London.

Hansbo S (1993) 'Band drains' in M P Moseley (ed.) *Ground Improvement*. Blackie Academic & Professional, Glasgow, pp. 40–64.

Hardisty P E, Croft R G, Bracken R A, Ross S D and Maccagno M D (1996) 'Design and application of horizontal wells for soil and groundwater remediation' in

M C Forde (ed.) *Proceedings of the Fourth International Conference on Polluted and Marginal Land*. Engineering Technical Press, Edinburgh, pp. 189–99.

Harnan C N and Iagolnitzer Y (1994) 'Colmix: the process and its applications' in A L Bell (ed.) *Grouting in the Ground*. Thomas Telford, London, pp. 511–24.

Harries C R, Witherington P J and McEntee J M (1995) *Interpreting Measurements of Gas in the Ground*, CIRIA Report 151. Construction Industry Research and Information Association, London.

Harris D I (2001) 'Protective measures' in J B Burland, J R Standing and F M Jardine (eds) *Building Response to Tunnelling: Case Studies from the Jubilee Line Extension, London*, CIRIA Special Publication 200. Thomas Telford, London.

Harris J S (1995) *Ground Freezing in Practice*. Thomas Telford, London.

Harris M R and Herbert S M (1994) *ICE Design and Practice Guide – Contaminated Land. Investigation, Assessment and Remediation*. Thomas Telford, London.

Harris M R, Herbert S M and Smith M A (1995) *Remedial Treatment for Contaminated Land*, Vol. III: *Site Investigation and Assessment*, CIRIA Report SP103. Construction Industry Research and Information Association, London.

Hayward D (2002) 'Dublin site yields buried treasure'. *Ground Engineering* 35(2): 30–1. Emap Construct, London.

Head K H (1986) *Manual of Soil Laboratory Testing*, Vols 1–3. Pentech Press, London.

Healy P R and Head J M (1984; reissued 2002) *Construction over Abandoned Mine Workings*, CIRIA Report SP32. Construction Industry Research and Information Association, London.

Henn R W (2001) *Backfill and Contact Grouting of Tunnels and Shafts*. Thomas Telford, London.

Herbst T F (1997) 'Removable ground anchors – answer for urban excavations' in G S Littlejohn (ed.) *Ground Anchorages and Anchored Structures*. Thomas Telford, London, pp. 197–205.

Higginbottom I E (1987) 'Methods of development above ancient shallow pillar-and-stall workings' in A K M Rainbow (ed.) *Reclamation, Treatment and Utilisation of Coal Mining Waste*. Elsevier, Amsterdam, pp. 639–52.

HMSO (1961) *The Construction (General Provisions) Regulations* (SI 1961/1580 with amendments) Her Majesty's Stationery Office, London.

HMSO (1988) *Form 91 Part 1 Section B Records of weekly thorough examination of excavations, cofferdams etc.* Her Majesty's Stationery Office, London.

HMSO (1990) Circular from Department of the Environment Transport and the Regions *'Environment Protection Act 1990: Part IIA Contaminated Land'*. Her Majesty's Stationery Office, London.

HMSO (1994) *The Construction (Design and Management) Regulations (1994) with Explanatory Note* (The 'CDM Regs') (SI 1994/3140 with amendments) Her Majesty's Stationery Office, London.

Hobson D M (1993) 'Rational site investigations' in T Cairney (ed.) *Contaminated Land, Problems and Solutions*. Chapman & Hall, London, pp. 29–67.

Hoek E and Brown E T (1980) *Underground Excavations in Rock*. Institution of Mining and Metallurgy, London.

Hoek E and Bray J W (1981) *Rock Slope Engineering*, 3rd edn. Spon Press, London.

Holden J T (1997) 'Improved thermal calculations for artificially frozen shaft excavations'. *Journal of Geotechnical and Geoenvironmental Engineering* (Aug.) 123(8): 696–701. American Society of Civil Engineers, New York.

Hope V (2003) 'Approach path'. *Ground Engineering* 36(10): 26–7. Emap Construct, London.

Houlsby A C (1982) 'Cement grouting for dams' in *Proceedings of Conference on Grouting in Geotechnical Engineering, New Orleans.* American Society of Civil Engineers, New York, pp. 1–34.

Houlsby A C (1985) 'Cement grouting: water minimising practices' in W Hayward Baker (ed.) *Issues in Dam Grouting.* Geotechnical Engineering Division,

American Society of Civil Engineers, New York, pp. 34–75.

Howden C and Crawley J D (1995) 'Sizewell B power station – design and construction of the diaphragm wall' in *Proceedings of Institution of Civil Engineers*, 108. London, pp. 48–62.

HSE (1985) *The Abbeystead Explosion Report*. Health and Safety Executive, Stationery Office, London.

HSE (1991) *Protection of workers and general public during the development of contaminated sites*. Health and Safety Executive, London.

HSE (1995) *The Construction (Design and Management) Regulations 1994. The roles of the client, planning supervisor, designer and the health and safety plan* (The CDM Regs) Health and Safety Executive, London.

HSE (1996) *A Guide to the Work in Compressed Air Regulations* (SI 1996/1656) Health and Safety Executive, London.

HSE (1997) *Dust general principles of protection*. Health and Safety Executive, London.

HSE (1998) *Guidance on the Noise at Work Regulations* (SI 1989: 1790) Health and Safety Executive, London.

HSE (1999) *Health and safety in excavations* 'Be safe and shore'. Health and Safety Executive, London.

HSE (2000) *A Guide to Risk Assessment Requirements*. Health and Safety Executive, London.

HSE (2002) *Control of Substances Hazardous to Health Regulations* ('COSSH Regs' amended by SI 2003/978) *'Approved code of practice and guidance' and 'Easy steps to control chemicals'*. Health and Safety Executive, London.

Hughes J M O and Withers N J (1974) 'Reinforcing of soft cohesive soils with stone columns'. *Ground Engineering* (May) 7(3): 42–9. Foundations Publications, London.

Humpherston C, Fitzpatrick A J and Anderson J M D (1986) 'The basement and substructure for the new headquarters of the Hong Kong and Shanghai Banking Corporation, Hong Kong' in *Proceedings of the Institution of Civil Engineers*, 80, Part 1. London, pp. 851–83.

Hutchinson J N (1984) 'Landslides in Britain' *Journal of Japanese Landslide Society* 21: 1–25.

ICE (1987) *Specification for Ground Treatment*. Institution of Civil Engineers, Thomas Telford, London.

ICE (1993) *Site Investigation in Construction* series. Vol. 1: *Without Site Investigation Ground is a Hazard*. Vol. 2: *Planning, Procurement and Quality Management*. Vol. 3: *Specification for Ground Investigation*. Vol. 4: *Guidelines for the Safe Investigation by Drilling of Landfills and Contaminated Land*. Institution of Civil Engineers, Thomas Telford, London.

ICE (1996a) *Specification for Piling and Embedded Retaining Walls*. Institution of Civil Engineers, Thomas Telford, London.

ICE (1996b) *Sprayed Concrete Linings (NATM) in Soft Ground*. Institution of Civil Engineers, Thomas Telford, London.

ICE (1999) *Specification for Construction of Slurry Trench Cut-off Walls as Barriers to Pollution Migration*. Institution of Civil Engineers, Thomas Telford, London.

ICE (2000) *Specification for Tunnelling* (in conjunction with the British Tunnelling Society). Institution of Civil Engineers, Thomas Telford, London.

ICE (2001) *Tunnel Lining Design Guide* (in conjunction with the British Tunnelling Society). Institution of Civil Engineers, Thomas Telford, London.

ICE (2002) *RAMP – Risk Analysis and Management for Projects*, rev. edn. Institution of Civil Engineers, Thomas Telford, London.

Jameson R, Pellegrino G and Shea M (1998) 'Instrumentation for performance monitoring of jet grouting' in *Proceedings of the Second International Conference on Ground Improvements*, CI Premier PTE Ltd, Singapore, pp. 613–20.

Jarvis S T and Brooks T G (1996) 'The use of flyash: cement pastes in the stabilisation of abandoned mineworkings'. *Waste Management* 16(1–3): 135–43. Elsevier, Science Ltd, London.

Jefferis S A (1985) 'Discussion on Arab Potash solar evaporation system' in *Proceedings of the Institution of Civil Engineers*, Part 1. London, pp. 641–6.

Jefferis S A (1993) 'In-ground barriers' in T Cairney (ed.) *Contaminated Land, Problems and Solutions*. Chapman & Hall, London, pp. 111–40.

Jewell P (2003) 'Anchor advances'. *Ground Engineering* 36(4): 33–5. Emap Construct, London.

Joyce M D (1982) *Site Investigation Practice*. Spon, London.

Karol R H (1990) *Chemical Grouting*. Marcel Decker Inc., New York.

Kennard M F and Reader R A (1975) 'Cow Green dam and reservoir' in *Proceedings of the Institution of Civil Engineers*, 58, Part 1. London, pp. 147–75.

King J C and Bindhoff E W (1982) 'Lifting and levelling heavy concrete structures' in *Proceedings of Conference on Grouting in Geotechnical Engineering, New Orleans*. American Society of Civil Engineers, New York, pp. 722–37.

Kutzner (1996) *Grouting in Soil and Rock*. A A Balkema, Rotterdam.

Lamont-Black J (2001) 'EKG: the next generation of geosynthetics' *Ground Engineering* 34(10): 22–3. Emap Construct, London.

Littlejohn G S (1982) 'Design of cement-based grouts' in *Proceedings of the Conference on Grouting in Geotechnical Engineering, New Orleans*. American Society of Civil Engineers, New York, pp. 35–48.

Littlejohn G S (1993) 'Chemical grouting' in M P Moseley (ed.) *Ground Improvement*. Blackie Academic & Professional, Glasgow, pp. 100–30.

Littlejohn G S and Waterhouse A (1990) 'Field data on long-range pumping of cement based grouts' in *Proceedings, Institution of Civil Engineers*, 88, Part 1. London, pp. 465–70.

Lord J A, Clayton C R I and Mortimore R N (2002) *Engineering in Chalk*, CIRIA Report C574. Construction Industry Research and Information Association, London.

Lunne T, Robertson P K and Powell J J M (1997) *Cone Penetrometer Testing in Geotechnical Practice*. Blackie Academic & Professional, London.

McDowell P W and Poulson A J (1996) 'Ground subsidence related to dissolution of chalk in Southern England'. *Ground Engineering* 29(3): 29–33. Emap Construct, London.

McDowell P W, Barker R D, Butcher A P, Culshaw M G, Jackson P D, McCann D M, Skipp B O, Matthews S L and Arthur J C R (2002) *Geophysics in Engineering Investigations*, Engineering Geology Special Publication 19 (also available as CIRIA Report C562). Construction Industry Research and Information Association, London.

McKenna J M, Horswill P and Smith E J (1985) 'Empingham Reservoir – seepage control measures' in *Proceedings of Institution of Engineers*, Part 1, 78 8827. London, pp. 219–46.

McNicholl D P, Pump W L and Cho G W F (1986) 'Groundwater control in large scale slope excavations – five case histories from Hong Kong' in J C Cripps, F G Bell and M G Culshaw (eds) *Groundwater in Engineering Geology*, Geological Society Engineering Geology Special Publication No. 3. The Geological Society of London, pp. 510–18.

McQuade S J and Needham A D (1999) 'Geomembrane liner defects – causes, frequency and avoidance' in *Proceedings of the Institution of Civil Engineers*, 137. London, pp. 203–13.

Maddison J D, Jones D B, Bell A L and Jenner C G (1996) 'Design of an embankment supported using low strength geogrids and vibro-concrete columns' in MB de Groot, G den Hoedt and R J Termaat (eds) *Geosynthetics Applications, Design and Construction, Proceedings of the First European Geosynthetics Conference*. Maastrict, A A Balkema, Rotterdam, pp. 325–32.

Masters-Williams H, Heap A, Ketts H, Greenshaw L, Davis S, Fisher P, Hendrie M and Owens D (2001) *Control of Pollution from Construction Sites: Guidance for Contractors*, CIRIA Report C532. Construction Industry Research and Information Association, Thomas Telford, London.

Medical Research Council (1982) *Medical Code of Practice for Work in Compressed Air*, CIRIA Report

R44. Construction Industry Research and Information Association, London.

Miller E (1988) 'The Eductor Dewatering System'. *Ground Engineering* 21(6): 29–34. Thomas Telford, London.

Mitchell J K and Katti R K (1981) 'Soil improvement: general report' in *Proceedings of the Tenth International Conference, Soil Mechanics and Foundation Engineering, Stockholm*, Vol. 4. A A Balkema, Rotterdam, pp. 567–78.

Mitchell J M and Jardine F M (2002) *A Guide to Ground Treatment*, CIRIA Report C573. Construction Industry Research and Information Association, London.

Mohamedelhassan E and Shang J Q (2001) 'Effects of electrode materials and current intermittance in electro-osmosis'. *Ground Improvement* 105(1): 3–11. Thomas Telford, London.

Monnet A and Iagolnitzer Y (1994) 'Analysis of a dynamic pumping test in an area enclosed by cut-off walls'. *Ground Engineering* 27(6): 25–33. Thomas Telford, London.

Moseley M P and Priebe H J (1993) 'Vibro techniques' in M P Moseley (ed.) *Ground Improvement*. Blackie Academic & Professional, Glasgow, pp. 1–19. (New edition in press by Spon Press, London.)

Myles B (1995) 'A rapid technique for soil nailing slopes' in T S Ingold (ed.) *The Practice of Soil Reinforcing in Europe*. Thomas Telford, London, pp. 241–52.

Nicholson D P, Gammage C and Chapman T (1994) 'The use of finite element methods to model compensation grouting' in A L Bell (ed.) *Grouting in the Ground*. Thomas Telford, London, pp. 297–312.

Nicholson D, Tse C-M and Penny C (1999) *The Observational Method in Ground Engineering – Principles and Applications*, CIRIA Report R185. Construction Industry Research and Information Association, London.

OCMA Specification (1973) *Code of Practice DFCP4 Drilling Fluid Materials: Bentonite*. Oil Companies Materials Association, London.

Parker P J, Wyllie M and Esnault A (1996) 'Investigation, planning & execution of the remediation of Ardeer landfill, Scotland' in M C Forde (ed.) *Polluted & Marginal Land – 96, Proceedings of the Fourth International Conference*. Engineering Technics Press, Edinburgh, pp. 153–68.

Peart R J, Cuss R J, Beamish D and Jones D G (2003) 'Eye in the sky'. *Geoscientist* 13(7): 4–7. Geological Society of London.

Peck R B, Hanson W E and Thornburn T H (1974) *Foundation Engineering*, 3rd edn. John Wiley & Sons Inc., New York.

Perry P (2002) *CDM Questions and Answers*. Thomas Telford, London.

Pine R J and Harrison J P (2003) 'Rock mass properties for engineering design'. *Quarterly Journal of Engineering Geology and Hydrogeology* 36(1): 5–16. Geological Society of London.

Porbaha A (1998) 'State of the art in deep mixing technology. Part I: Basic concepts and overview'. *Ground Improvement* 102(2): 81–92. Thomas Telford, London.

Porbaha A (2000) 'State of the art in deep mixing technology. Part IV: Design considerations'. *Ground Improvement* 104(3): 111–25. Thomas Telford, London.

Porbaha A, Tanaka M and Kobayashi M (1998) 'State of the art in deep mixing technology. Part II: Applications'. *Ground Improvement* 102(3): 125–39. Thomas Telford, London.

Porbaha A, Shibuya S and Kishida T (2000) 'State of the art in deep mixing technology. Part III: Geomaterial characterisation'. *Ground Improvement* 104(3): 91–110. Thomas Telford, London.

Powell J H (1998) *A Guide to British Stratigraphic Nomenclature*, CIRIA Special Publication 149. Construction Industry Research and Information Association, London.

Powers J P (1992) *Construction Dewatering – New Methods and Applications*. John Wiley & Sons Inc., New York.

Powrie W and Preene M (1994) 'Performance of ejectors in construction dewatering systems'. *Geotechnical Engineering* 107: 143–54. Institution of Civil Engineers, London.

Powrie W, Roberts T O L and Jefferis S A (1990) 'Biofouling of site dewatering systems' in P Howsam (ed.) *Microbiology in Civil Engineering*. E & F N Spon, London.

Powrie W, White J K and Preene M (1993) *The Design of Pore Pressure Control Systems in Fine Soils*, CIRIA Project Report 10. Construction Industry Research and Information Association, London.

Preene M (2000) 'Assessment of settlements caused by groundwater control'. *Geotechnical Engineering* 143(4): 177–90. Institution of Civil Engineers. London.

Preene M, Roberts T O L, Powrie W and Dyer M R (2000) *Groundwater Control – Design and Practice*, CIRIA Report C515, Construction Industry Research and Information Association, London.

Price D G and Plaisted A C (1971) 'Epoxy resins in rock slope stabilisation works' in *'Sympsosium, Société International Mécanique des Roches*, Nancy, III-9.

Price D G, Malkin A B and Krill J L (1969) 'Foundations of multi-storey blocks on the Coal Measures with special reference to old mine workings'. *Quarterly Journal of Engineering Geology*, 1(4): 271–322. Geological Society of London.

Priebe H J (1995) 'The design of vibro-replacement'. *Ground Engineering* 28(10): 31–7. Thomas Telford, London.

Priebe H J (1998) 'Vibro-replacement to prevent earthquake induced liquefaction'. *Ground Engineering* 31(9): 30–3. Emap Construct, London.

Privett K D, Matthews S C and Hodges R A (1996) *Barriers, Liners and Cover Systems for Containment and Control of Land Contamination*, CIRIA Special Publication 124. Construction Industry Research and Information Association, London.

Puller M (1996) *Deep Excavations. A Practical Manual*. Thomas Telford, London.

Raabe E W and Esters K (1990) 'Soil fracturing techniques for terminating settlements and restoring levels of building and structures'. *Ground Engineering* 23(4): 33–45. Thomas Telford, London.

Raffle J F and Greenwood D A (1961) 'The relationship between the rheological characteristics of grouts and their capacity to permeate soils' in *Proceedings of the Fifth International Conference on Soil Mechanics and Foundation Engineering, Dunod, Paris*. Vol. II, pp. 789–93.

Rawlings C G, Hellawell E E and Kilkenny W M (2000) *Grouting for Ground Engineering*, CIRIA Report C514, Construction Industry Research and Information Association, London.

RDGC (1993) *Soil Nailing Recommendations – Clouterre Program (1991)*. Projets Nationaux de Recherche Développement en Génie Civil, Presses de l'EPNC, Paris. (English version published by Federal Highway Administration, Baltimore, MD.)

Rice S (2002) 'Clearing the air'. *Ground Engineering* 35(11): 36–7. Emap Construct, London.

Rogers C D F, Glendinning S and Dixon N (eds) (2001) *Lime Stabilisation*. Thomas Telford, London.

Schlosser F, Unterreiner P and Pumelle C (1992) 'French research programme Clouterre on soil nailing' in *Proceedings of the Conference on Grouting and Soil Improvement, New Orleans*. American Society of Civil Engineers, New York, pp. 172–86.

Sheen M S (2001–2) 'Grouting pressure and damaged adjacent buildings. Part I: Behaviour analysis' and 'Part II: Case Study', *Ground Improvement* 5(4): 155–62 and 6(2): 47–58. Thomas Telford, London.

Sherwood D E, Haman C N and Beyer M G (1989) 'Recent developments in secant bored pile wall construction' in *Proceedings of the International Conference on Piling and Deep Foundations*. A A Balkema, Rotterdam, pp. 211–20.

Shields J G (1997) 'Post-tensioning Mullardoch Dam in Scotland' in G S Littejohn (ed.) *Ground Anchorages and Anchored Structures*. Thomas Telford, London, pp. 205–16.

Simpson B and Driscoll R (1998) *Eurocode 7 – A Commentary*. Building Research Establishment, Watford and Construction Research Communications Ltd, London.

Simpson B, Blower T, Craig R N and Wilkinson W B (1989) *The Engineering Implications of Rising Groundwater Levels in the Deep Aquifer beneath London*, CIRIA Report SP69. Construction Industry Research and Information Association, London.

Skempton A W (1977) 'Slope stability of cuttings in brown London clay' in *Proceedings of the Ninth International Conference on Soil Mechanics,* Japanese Society of Soil Mechanics and Foundation Engineering, Tokyo, 3, pp. 261–70.

Skempton A W and Coates D J (1985) 'Carsington Dam failure' in *Failures in Earthworks*. Thomas Telford, London, pp. 203–20.

Slocombe B C (1993) 'Dynamic compaction' in M P Moseley (ed.) *Ground Improvement*. Blackie Academic & Professional, Glasgow, pp. 20–39. (New edition in press by Spon Press, London.)

Somerville S H (1986) *Control of Groundwater for Temporary Works*, CIRIA Report 113. Construction Industry Research and Information Association, London.

Soudain M (2003) 'Multiple choice'. *Ground Engineering* 36(2): 29–30. Emap Construct, London.

Statham I and Scott M (1994) 'An analysis of drill and grout records from the South Wales coalfields' in A L Bell (ed.) *Grouting in the Ground*. Thomas Telford, London, pp. 103–24.

Tomlinson M J (1994) *Pile Design and Construction Practice,* 4th edn. E & F N Spon, London.

Tomlinson M J (2001) *Foundation Design and Construction,* 7th edn. Prentice-Hall, Harlow, Essex.

Tomlinson M J and Wilson D M (1973) 'Preloading of foundations by surcharge on filled ground'. *Geotechnique* 23(1): 117–20. Institution of Civil Engineers, London.

Trenter N (2003) 'Understanding and containing geotechnical risk' in *Proceedings of Institution of Civil Engineers*, 156(1). London, pp. 42–8.

TSO (1996a) *The Construction (Health, Safety and Welfare) Regulations* (SI 1996/1592 with amendments) The Stationery Office, London.

TSO (1996b) *Offshore installations and wells (Design and construction etc) Regulations* (SI 1996/913) The Stationery Office, London.

TSO (2002) *Control of Noise (Code of Practice for Construction and Open Sites – England)* (SI 2002/461 etc) The Stationery Office, London.

Tucker M E (1996) *Sedimentary Rocks in the Field*. Geological Society of London Field Guide Series, Wiley, Chichester.

Turner M J (1997) *Integrity Testing in Piling Practice*, CIRIA Report 144. Construction Industry Research and Information Association, Institution of Civil Engineers, London.

van der Stoel A E C (2001) *Grouting for Pile Foundation Improvement*. DUP Science, Delft.

Waltham A C (1994) *Foundations of Engineering Geology*. Spon, London.

Waltham A C and Fookes P G (2003) 'Engineering classification of karst ground conditions'. *Quarterly Journal of Engineering Geology and Hydrogeology* 36(2): 101–18. Geological Society of London.

Warren C D and Mortimore R N (2003) 'Chalk engineering geology – Channel Tunnel Rail Link and North Downs Tunnel'. *Quarterly Journal of Engineering Geology and Hydrogeology* 36(1): 17–34. Geological Society of London.

Watson P (2003) 'Channel Tunnel Rail Link, section 1: North Downs tunnel' in *Proceedings of Institution of Civil Engineers*, 156. London, pp. 40–8.

Watts K S (2000) *Specifying Vibro Stone Columns*, Building Research Establishment Publication BR391. CRC Ltd, London.

White W B (1990) 'Surface and near-surface karst landforms' in C G Higgins and D R Coates (eds) *Groundwater Geomorphology: The Role of Subsurface Water in Earth-Surface Processes and Landforms*, Special Paper 252. Geological Society of America. Boulder, Colorado, pp. 157–75.

Whitlow R (1995) *Basic Soil Mechanics*, 3rd edn. Longman, Harlow, Essex. (New edition in press by Pearson Education, Harlow, Essex.)

Whitman R V (2000) 'Organizing and evaluating uncertainty in geotechnical engineering'. *Journal of Geotechnical and Geoenvironmental Engineering* 126(7): 583–93. American Society of Civil Engineers, New York.

Williams B P and Waite D (1993) *The Design and Construction of Sheet-Piled Cofferdams*, CIRIA Report SP95. Construction Industry Research and Information Association, London.

Wilson S A and Shuttleworth A (2002) 'Design and performance of a passive dilution gas migration barrier'. *Ground Engineering* 35(1): 34–7. Emap Construct, London.

Yeats J A and O'Riordan N J (1989) 'The design and construction of large diameter base grouted piles in Thanet Sand at Blackwall Yard, London' in *Proceedings of the International Conference on Piling and Deep Foundations*. A A Balkema, Rotterdam, pp. 455–61.

INDEX